Innovative Subsurface Remediation

ACS SYMPOSIUM SERIES **725**

Innovative Subsurface Remediation

Field Testing of Physical, Chemical, and Characterization Technologies

Mark L. Brusseau, EDITOR
University of Arizona

David A. Sabatini, EDITOR
University of Oklahoma

John S. Gierke, EDITOR
Michigan Technical University

Michael D. Annable, EDITOR
University of Florida

American Chemical Society, Washington, DC

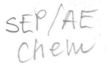

Library of Congress Cataloging-in-Publication Data

Innovative subsurface remediation : field testing of physical, chemical, and characterization technologies / Mark L. Brusseau, editor ... [et al.].

p.   cm.—(ACS symposium series ; 725)

Includes bibliographical references and index.

ISBN 0-8412-3596-1 (cl.)

1. Soil remediation—Technological innovations—Evaluation. 2. Hazardous waste site remediation—Technological innovations—Evaluation.

I. Brusseau, Mark L. II. Series.

TD878.I554   1999
628.5'5—dc21   99-12285
                                                                      CIP

The paper used in this publication meets the minimum requirements of American National Standard for Information Sciences—Permanence of Paper for Printed Library Materials. ANSI Z39.48–1984.

Foreword

THE ACS SYMPOSIUM SERIES was first published in 1974 to provide a mechanism for publishing symposia quickly in book form. The purpose of the series is to publish timely, comprehensive books developed from ACS sponsored symposia based on current scientific research. Occasionally, books are developed from symposia sponsored by other organizations when the topic is of keen interest to the chemistry audience.

Before agreeing to publish a book, the proposed table of contents is reviewed for appropriate and comprehensive coverage and for interest to the audience. Some papers may be excluded in order to better focus the book; others may be added to provide comprehensiveness. When appropriate, overview or introductory chapters are added. Drafts of chapters are peer-reviewed prior to final acceptance or rejection, and manuscripts are prepared in camera-ready format.

As a rule, only original research papers and original review papers are included in the volumes. Verbatim reproductions of previously published papers are not accepted.

ACS BOOKS DEPARTMENT

Contents

INTRODUCTION

INNOVATIVE IN SITU PHYSICAL
AND CHEMICAL REMEDIATION TECHNOLOGIES

INNOVATIVE SITE CHARACTERIZATION METHODS

INDEXES

Preface

The clean-up of subsurface contamination is a daunting enterprise, constrained by numerous factors. It is now recognized that the remediation technologies currently in widespread use (e.g., pump and treat, soil venting), cannot solve all subsurface contamination problems. Furthermore, it is clear that the selection, design, implementation, and evaluation of a remediation system cannot be optimized without accurate and sufficient site characterization. This recognition has fueled the development of new, so-called "innovative" remediation and characterization technologies. Many of these innovative technologies are in fact not new at all. For example, surfactant flushing has a long history of use in the petroleum field. However, its use in the environmental field is more recent and faces unique challenges, hence its "innovative" status. In any event, new technologies continue to be developed, increasing our arsenal of potential tools available for subsurface characterization and remediation.

There is, however, an enormous gulf between the development of a technology and its widespread use in the field. Barriers to using innovative technologies are many, including technical-related factors, regulatory obstacles, conservatism on the part of applicators–users, and the lack of reliable performance data. This last factor, "proof of performance", is central to the acceptance and use of any innovative technology. Definitive proof of performance can be obtained only from field tests, conducted under controlled, well-characterized conditions. Interest in the plannning, implementation, and evaluation of such tests has greatly expanded among researchers, applicators, and regulators.

This book arises from our interest in this topic, which led us to organize this special symposium, *Innovative Subsurface Remediation: Field Testing of Physical, Chemical, and Characterization Technologies*. The results of field tests of several innovative remediation technologies are reported herein. The results of the tests are discussed, along with the trials and tribulations associated with the tests. In addition, overviews and examples of innovative subsurface characterization methods are presented, with a focus on tracer-based methods. This information should prove useful to those interested in the remediation and characterization of contaminated soil and groundwater.

We thank the authors for their participation in this endeavor; without them there would be no book. We thank the many reviewers who helped improve the content of the chapters. Finally, we thank the staff of the ACS Books Department, particularly Thayer Long, Maureen Matkovich, David Orloff, and Anne Wilson, for their critical and timely assistance.

MARK L. BRUSSEAU
Department of Soil, Water, and Environmental Science
  and Department of Hydrology and Water Resources
University of Arizona
Tucson, AZ 85721

DAVID A. SABATINI
Department of Civil and Environmental Engineering
University of Oklahoma
Norman, OK 73019–0631

JOHN S. GIERKE
Department of Geological Engineering and Sciences
Michigan Technical University
Houghton, MI 49931–1295

MICHAEL D. ANNABLE
Department of Environmental Engineering Sciences
Black Hall University
University of Florida
Gainesville, FL 32611

INTRODUCTION

Chapter 1

Field Demonstrations of Innovative Subsurface Remediation and Characterization Technologies: Introduction

Mark L. Brusseau[1], John S. Gierke[2], and David A. Sabatini[3]

[1]Department of Soil, Water, and Environmental Science and Department of Hydrology and Water Resources, University of Arizona, Tucson, AZ 85721
[2]Department of Geological Engineering and Sciences, Michigan Technological University, Houghton, MI 49931–1295
[3]Department of Civil Engineering and Environmental Science, University of Oklahoma, Norman, OK 73019

Groundwater pollution has become one of our most pervasive environmental problems, and remediating sites with contaminated groundwater has proven to be a formidable challenge. Remediation efforts are often limited by the complexity of the subsurface environment, and by our lack of knowledge of that environment. Recent research has focused on enhancing our understanding of the complex subsurface environment and on developing innovative technologies capable of handling these complexities. An important step in the evolution of a new technology is going beyond well-controlled laboratory experiments to testing the technology at the field scale (i.e., real world). The purpose of this volume is to present evaluations of selected innovative technologies that have undergone field demonstration testing. This volume also reports on recent advances in subsurface characterization techniques that are critical to the proper design of all technologies, and that can help assess the performance of these technologies.

The Need for Innovative Technologies

Initial efforts to remediate groundwater contamination focused on pump-and-treat approaches (flushing water through the formation and treating the extracted fluids). This approach has proven to be generally successful for preventing further spread of contamination (plume containment). Unfortunately, it has been generally ineffective for restoring systems to "pristine" conditions (1-3). Numerous factors are responsible for the ineffectiveness of pump-and-treat remediation, with one of the most common being zones of high concentrations or large masses of contaminants (i.e., source zones). While pump and treat can effectively manage the dilute portion of the plume, contaminant mass removal from the source zones is usually limited by equilibrium (e.g., solubility) and kinetic (e.g., dissolution, desorption) related constraints. For example, as noted in a recent National Research Council report (3), the presence of immiscible-liquid contamination in the subsurface is often the single most critical factor influencing site remediation. Successful remediation of many contaminated subsurface systems is dependent on the development and implementation of technologies that can control or remove source-zone contamination.

If source zones are present, the associated groundwater contaminant plumes can not be remediated effectively unless the source zones are at least controlled. This can be accomplished in a number of ways. For example, physical barriers (e.g., slurry walls) can be used to isolate the source zone, thus shutting off the supply of contamination to the plume. Hydraulic controls can also be used, especially for deep or large source zones where physical barriers are impractical. Of course, the control system will need to be managed "in perpetuity", which in itself may be costly.

A different approach to source control involves altering the chemical nature of the contaminant to reduce its mobility. This may be accomplished, for example, by injecting a reagent that promotes binding of the contaminant to the solid phase of the porous medium. A major technical and regulatory concern for this approach is the potential reversibility of the treatment.

An alternative to source control is actually reducing the contaminant mass resident in the source zone. This can be accomplished by either increasing the mass removal rate associated with pump and treat, or by promoting in-situ chemical or biological transformation reactions. Many of the innovative technologies currently in development are focused on source-zone remediation.

Introducing chemical amendments to enhance pump-and-treat removal of organic and inorganic compounds is discussed by Palmer and Fish (4). Chemical amendments include complexing agents, cosolvents, surfactants, oxidation-reduction agents, precipitation-dissolution reagents, and ionization reagents. The use of surfactants and cosolvents for enhanced removal of immiscible-liquid contamination has become a major focus of research, and is considered to hold promise for improving site remediation (5). Thus, these technologies are discussed in detail in this volume.

Promoting in-situ biotransformation of contaminants is another approach receiving enormous attention. Methods based on this approach are covered extensively in other venues and are therefore not discussed herein. The promotion of in-situ chemical transformations has received much less attention. One example of a demonstration of this approach is covered in this volume.

It is important to stress that many of these technologies are designed for implementation in source zones, which are generally much smaller in size compared to contaminant plumes. In addition, many of the technologies are based on the injection of amendments, which may involve a relatively large materials cost. As such, implementation of these technologies can not be viewed in the same manner as traditional technologies. For example, the large materials cost associated with a surfactant flush may be more than compensated by the cost savings associated with the elimination of many years of pump-and-treat operation. Thus, it is important to do a comprehensive cost-benefit analysis as part of the technology evaluation process.

The heterogeneous nature of the subsurface is another major constraint to the successful cleanup of contaminated subsurface environments. Clearly, the selection, design, implementation, and evaluation of a remediation system can not be optimized without accurate and sufficient information on the physical, chemical, and biological properties of the subsurface. This includes characterizing (1) the magnitude and variability of important material properties (e.g., hydraulic conductivity, bacterial populations, pH), (2) the type, amount, and distribution of contaminants, and (3) the major contaminant transport and fate processes. Currently, most sites are not characterized sufficiently due to cost and time constraints. The development of new site-characterization technologies will enhance the effectiveness of current remediation technologies and provide the basis for implementation of new innovative technologies.

Most current characterization methods are based on "point sampling": groundwater monitoring wells for aqueous samples and coring for solid-phase sampling. Newer technologies based on geoprobes are also point sampling methods. These methods have the potential to provide accurate and precise data for a small domain. This is an advantage for certain applications. However, point-sampling methods are also constrained by the size limitation. Specifically, because of the heterogeneity inherent to the subsurface, it is very difficult to accurately characterize a system without a cost- and time-prohibitive number of sampling points. In addition, the use of geostatistical methods for calculating representative values for the sampled domain is rarely straightforward. Thus, methods that provide measurements over much larger areas are being developed. One such group of technologies is based on the use of tracers. Several examples of "innovative tracer tests" are presented in this volume.

The Need for Field Demonstrations of Innovative Technologies

Many groundwater hydrologists and environmental engineers initially thought that groundwater remediation would be relatively simple. Groundwater hydrologists were routinely analyzing well fields, while environmental engineers were adept at designing waste water treatment plants. It seemed that together these two disciplines should be able to design effective groundwater remediation systems. However, it quickly became evident that issues relatively unimportant to groundwater flow were critically important to contaminant transport, and that the subsurface environment is dramatically different from the "ideal" reactors used in industrial facilities. The complexity of the subsurface environment thus proved to have much more of an impact on remediation technology performance than either discipline anticipated.

Subsurface complexities also pose a formidable challenge to the feasibility and effectiveness of innovative remediation technologies. As a result, many technologies that performed wonderfully in the laboratory may fail in the field. The key question for any innovative technology is, of course, will that technology perform at the field scale? Clearly then, "proof of performance" is central to the acceptance and use of any remediation technology. Unfortunately, however, proof of performance is lacking for many innovative technologies. There are three primary reasons why proof of performance may be lacking for a given technology: (1) field-scale performance tests have not been conducted, (2) the available performance data are poor due to poorly conducted tests, (3) the results of properly conducted tests are not disseminated to interested parties.

The first two factors can be addressed by conducting proper performance tests for the specific technology. A proper performance test should answer these two questions (6): (1) does the technology reduce the risks posed by the contamination (i.e., reduce mass, mobility, toxicity)?; (2) was the technology the cause of the risk reduction? The most important element of an effective performance assessment is a carefully controlled field test of the technology (6). This test should include a comprehensive site characterization and the development of a sound understanding of the factors controlling contaminant transport and fate (7,8).

For the third constraint, dissemination of results, Dzombak (9) and Gierke and Powers (7) suggest that the publication venue is a critical component of acceptance. Comprehensive works appearing in refereed literature were found to be the most useful. Conversely, bibliographies, literature reviews and summaries, conference proceedings, newsletters, and database software appeared to be of limited value in selecting and designing innovative treatment systems.

The studies presented in this volume were designed to provide an accurate and effective performance evaluation of their respective technologies. Thus, it is hoped that this information will be interesting and useful to those involved in subsurface remediation. Furthermore, we hope that this volume proves to be a successful method of disseminating this information.

References

1. Keeley, J. 1989. Performance Evaluations of Pump and Treat Remediations. USEPA, EPA/540/4-89-005. 19 pp.
2. Haley, J. L., Hanson, B., Enfield, C., and Glass, J. Ground Water Monitoring Review. Winter 1991, 119-124.
3. National Resource Council. **Alternatives for Groundwater Cleanup**; National Academy Press: Washington, D.C., 1994, 315 pp.
4. Palmer, C. D. and Fish, W. Chemical Enhancements to Pump and Treat Remediation. USEPA, EPA/540/S-92/001, 1992, 20 pp.
5. USEPA (U.S. Environmental Protection Agency). 1992. Dense Nonaqueous Phase Liquids-- A Workshop Summary. USEPA, EPA/600/R-92/030.
6. National Research Council. **Innovations in Ground Water and Soil Cleanup**; National Academy Press: Washington, D.C., 1997, 292 pp.
7. Gierke, J. S.; Powers, S. E. *Water Environ. Res.* 1997, 69, 196.
8. Brusseau, M.L.; Rohrer, J.W.; Decker, T.M.; Nelson, N.T.;Linderfelt, W.R. 1998, in this volume.
9. Dzombak, D.A. *Water Environ. Res.* 1994, 66, 187.

Innovative In Situ Physical and Chemical Remediation Technologies

Chapter 2

Surfactant Selection Criteria for Enhanced Subsurface Remediation

David A. Sabatini[1], Jeffrey H. Harwell[2], and Robert C. Knox[1]

[1]The Schools of Civil Engineering and Environmental Science and
[2]Chemical Engineering and Materials Science, and The Institute for Applied
Surfactant Research, University of Oklahoma, Norman, OK 73019

Successful implementation of surfactant enhanced subsurface remediation requires careful consideration of fundamental surfactant properties, as highlighted in this chapter. The economic viability of this technology requires targeting the residual contamination, minimizing surfactant losses, and recovery and reuse of the surfactant, as discussed in this chapter. The relative efficiencies and advantages of surfactant solubilization and mobilization systems are described, as well as means to optimize these systems. Unit processes for contaminant-surfactant separation and surfactant reuse are summarized, along with unique surfactant impacts on these separation processes. Finally, factors affecting the environmental acceptability of surfactants are discussed. This chapter thus provides an introduction to surfactant chemistry and highlights key factors critical to the successful design and implementation of surfactant enhanced subsurface remediation systems.

Surfactant enhanced subsurface remediation is being widely considered as a method for enhancing pump and treat remediation. This technology takes advantage of unique surfactant properties, which are largely due to the two distinct regions of surfactant molecules; i.e., hydrophobic and hydrophilic moieties. Being amphiphilic, surfactants migrate to interfaces where both moieties exist in a preferred phase (e.g., air-water and oil-water interfaces). This accumulation alters interfacial properties, such as air-water or oil-water interfacial tension, and can also greatly increase the apparent contaminant solubility, as discussed below.

Surfactant Fundamentals

Surfactants (surface-active-agents) are classified by their charge (cationic, anionic, nonionic, zwitterionic -- having both cationic and anionic groups), their origin

8

(biosurfactants from plant or microbial production versus synthetic surfactants), their regulatory status (direct or indirect food additive status, acceptable for discharge to wastewater treatment systems or for use in pesticide formulations), etc. Surfactants are also characterized by their hydrophilic-lipophilic balance (HLB); surfactants with a high HLB value are water soluble while oil soluble surfactants have low HLB values (1).

Above a critical concentration surfactant monomers self-aggregate into micelles (Figure 1). The surfactant concentration above which micelles form is known as the critical micelle concentration (CMC). The hydrophilic micelle exterior makes them highly soluble in water. Having a hydrophobic (oil like) interior, micelles are sometimes described as dispersed oil drops. When surfactant concentrations exceed the CMC, the incremental surfactant goes almost totally to formation of additional micelles (i.e., the monomer concentration is virtually constant above the CMC). In micelles, the surfactant hydrophobe is in the interior of the aggregate while the hydrophilic moiety is on the micelle exterior (Figure 1).

Surfactant enhanced subsurface remediation is often categorized into two systems; solubilization and mobilization. Solubilization enhancement results from contaminant partitioning into the oil-like core of the micelle, thereby effectively increasing the aqueous solubility of the contaminant. With increasing solubility, the number of water flushings required to achieve a treatment goal decreases. At low surfactant concentrations (sub-CMC) the contaminant is present at its water solubility. Just above the CMC the solubility enhancement is minor, but increases as the surfactant concentration continues to increase above the CMC. The higher the surfactant concentration is above the CMC, the greater the number of micelles and thus the greater the solubility enhancement (as shown in Figure 2). Thus, to dramatically improve contaminant extraction we will operate significantly above the CMC (an order of magnitude or more).

As demonstrated in Figure 2, solubility enhancement is a function of the contaminant hydrophobicity. Solubilization enhancement is typically described by the molar solubilization ratio (MSR) and the micelle-water partition coefficient (K_m). The molar solubilization ratio is the moles of contaminant solubilized per mole of surfactant, and can be determined directly from a plot of contaminant solubility versus surfactant concentration (i.e., the slope above the CMC). The micelle-water partition coefficient (K_m) is the molar distribution of the contaminant between the micellar phase (X_m) and the aqueous phase (X_a), and can be determined from values of MSR, water solubility of the contaminant (S), and molar concentration of water (C_o) as follows (2, 3):

$$K_m = \frac{X_m}{X_a} = \frac{C_o}{S} \left[\frac{MSR}{1 + MSR} \right]$$

The data in Figure 2 and Table I illustrate that solubility enhancements (K_m values) for PCE are roughly an order of magnitude greater than TCE, while TCE enhancements are roughly an order of magnitude greater than DCE (3).

Mobilization refers to bulk oil displacement from the trapped residual oil (as opposed to simply enhancing the solubility). The bulk displacement occurs due to reductions in interfacial tension between the NAPL and water phases. It is this interfacial tension which is largely responsible for the NAPL trapping (residual saturation) in the

10

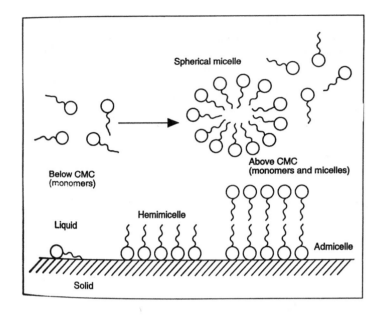

Figure 1: Examples of Surfactant Monomers, Micelles, Hemimicelles and Admicelles (Adapted from 9).

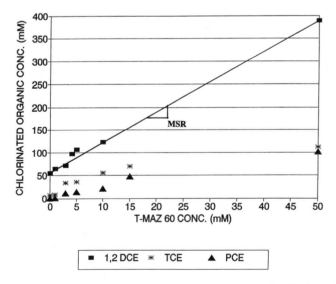

Figure 2: Example of Surfactant Enhanced Solubilization of Chlorinated Organic Compounds; Intercept is Aqueous Contaminant Concentration; MSR -- Molar Solubilization Ratio; (Adapted from 22).

Table I: Solubilization Parameters for Chlorinated Organic Compounds (3)

Chlorinated Organic	Surfactant	MSR	log K_m
PCE	SDS	0.39	4.5
	T-MAZ 28	0.45	4.55
	T-MAZ 20	2.27	4.9
	T-MAZ 60	3.15	4.94
TCE	SDS	0.34	3.27
	T-MAZ 28	1.68	3.66
	T-MAZ 20	3.29	3.75
	T-MAZ 60	3.95	3.77
1,2-DCE	SDS	1.37	2.76
	T-MAZ 28	2.46	2.85
	T-MAZ 20	7.49	2.95
	T-MAZ 60	6.91	2.94

porous media.   Significant reductions in the oil-water interfacial tension virtually eliminates the capillary forces which cause the oil to be trapped, thereby allowing the oil to readily flush out with the water. Thus, mobilization of the NAPL is maximized when ultra-low interfacial tensions are achieved. The minimum interfacial tension occurs in middle phase microemulsion systems. By adjusting the surfactant system it is possible to transition from normal micelles (aqueous phase, Winsor Type I), to middle phase microemulsions (Winsor Type III), to reverse micelles (oil phase inverted micelles, Winsor Type II). Careful system design will minimize the formation of mesophases in the transition region (e.g. liquid crystals), which may occur instead of the desired middle phase system. Traditionally alcohols have served this purpose (4), while more recently hydrotropes (e.g., naphthalene sulfonates) and cosurfactants (e.g., a second surfactant) have been evaluated.

Figures 3 and 4 illustrate the Winsor regions for a water-oil system containing equal volumes of water and oil. For a high HLB surfactant systems (right side of figure) the surfactant resides in the water phase as normal micelles and a portion of the oil phase partitions into the micellar phase. For low HLB surfactant systems (the left side of the figure) the surfactant resides in the oil phase as reverse micelles. In between we observe three phases; the water and oil phases and a new "middle" phase (so designated because of its intermediate density). The middle phase is comprised of water, oil and a majority of the surfactant. The interfacial tension reaches a minimum within this middle phase region (i.e., when equal volumes of oil and water occur in the middle phase system),

12

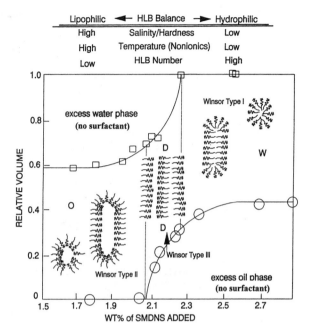

Figure 3: Surfactant-Oil Phase Behavior Diagram (1,2-DCE, 0.5 wt% AOT and scanning with Sodium Mono- and Di-Methyl Naphthalene Sulfonate); W: surfactant-water phase; O: surfactant-oil phase; D: middle phase); (Adapted from *22*).

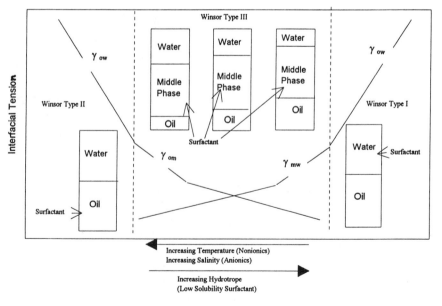

Figure 4: Interfacial Tension versus Surfactant Phase Diagram

thereby significantly enhancing the oil elution from the system (see Figure 4). Middle phase microemulsions also have ultra-high solubilization potentials, which can further enhance contaminant elution. Actually, solubilization potential and interfacial tension are inversely proportional; ultra-high solubilization systems also have ultra-low interfacial tension (4).

Middle phase microemulsions can be formed in several ways, as denoted at the top of Figure 3. For ionic surfactants, increasing salinity or hardness can produce the middle phase system, a strategy commonly utilized in surfactant enhanced oil recovery (5). However, introduction of high salt concentrations may not be desirable in aquifer restoration. Middle phase systems can also be achieved by altering the hydrophilic - lipophilic balance (HLB) of a binary surfactant system; recall that higher HLB values indicate increasing water solubility of the surfactant(s). By varying the mass ratio of two surfactants, one with an HLB above and one below the desired level, it is possible to achieve the desired HLB value.

As an example, a middle phase microemulsion system was achieved for 1,2-DCE by holding the AOT concentration constant and varying the SMDNS concentration (Figure 3). At low SMDNS concentrations AOT partitions into the oil phase and a Type II system is realized. With increased SMDNS concentration AOT partitions less into the oil phase, thereby allowing it to accumulate at the interface between the oil and water. At just the right combination the system HLB is balanced and a middle phase system results. As the SMDNS concentration is further increased, the system is over-optimized and the surfactants reside in the water phase (Type I system). This is the inverse of a salinity scan, where increasing salt concentrations causes a high-HLB, water-soluble surfactant to partition into the oil phase.

Recent studies in surfactant-enhanced environmental remediation, along with prior research on surfactant enhanced oil recovery (SEOR), have helped advance environmental remediation technologies; (e.g., 2, 5-14). Some important considerations from these and other references are provided below.

Economic Viability

Concluding that surfactants can be used to reduce remediation times does not necessarily demonstrate that surfactant use will be economical. In looking at two cases, Krebs-Yuill et al. (15) determined that surfactant solubilization can be more economical than conventional pump-and-treat. In the first case, sites from 1/4 acre to 10 acres were considered, with co-contamination by PCE, TCE, and 1,2-DCE, and assuming a remediation goal of seven years. For these conditions the two most important conclusions were: (1) surfactant capital costs constitute the single largest cost in a surfactant-enhanced remediation process, and (2) decontamination of the surfactant-stream and surfactant reuse is critical. The surfactant cost alone exceeded the cost of 25 years of pump-and-treat on a discounted cash flow basis for sites of 10 acres. Surfactant recovery costs were estimated to be less than 10% of the annual surfactant replacement costs. For sites of 1 acre or less, significant reductions in the total project costs were realized and surfactant enhanced remediation was competitive with water-based pump-and-treat remediation.

A subsequent economic study, in conjunction with DuPont, evaluated surfactant-enhanced pump-and-treat for residual zones of 2 acres or less (16). In this study it was

concerns, the potential efficiencies of mobilization systems cannot be dismissed. In fact, vertical circulation wells, neutral buoyancy systems and other technologies can potentially mitigate, and even take advantage of, vertical migration tendencies.

The Capillary Number can be used to assess the extent interfacial tension reduction necessary to displace residual NAPL. The Capillary Number (N_c) is the ratio of forces causing residual displacement to those resisting displacement (i.e., ratio of viscous to capillary forces), and can be defined as follows:

$$N_c = \frac{k \; \rho_w \; g \; \Delta H}{\sigma}$$

where k is the permeability, ρ_w is the density of water, g is the gravitational constant, ΔH is the hydraulic gradient, and σ is the NAPL-water interfacial tension. Typically, three to four orders of magnitude reduction in interfacial tension are required to displace residual oil. This depends on the capillary curve, which is a function of the contaminant-media system. Figure 5 shows residual oil saturation as a function of the capillary number (*4, 30*). In Figure 5, as the interfacial tension decreases the capillary number increases and additional oil is displaced, (i.e., the residual oil saturation declines). Figure 5 also shows Winsor Type I and Type III regions on a capillary number curve.

Another important parameter is the Bond Number (N_b), which is the ratio of buoyancy forces to capillary forces. This is defined as follows:

$$N_b = \frac{g \; k \; (\rho_w - \rho_o)}{\sigma}$$

where ρ_o is the density of the oil, and all other terms are as defined above. The Bond number is especially important for fluids with densities dramatically different from water (for example, PCE with a specific gravity of 1.6 would have a negative Bond Number).

Figure 6 shows results of one-dimensional column studies for solubilization and mobilization, and demonstrates that mobilization achieved higher concentrations and eluted the PCE more quickly than solubilization (> 97% extracted via mobilization in ca. 3 pore volumes -- (*29*). The lower maximum concentration, plateau and tail on the solubilization curve indicates the reduced extraction rate and thus slow approach to complete PCE elution for solubilization (likely due to interfacial area constraints); while ca. 85% of the PCE is eluted within 10 pore volumes, less than 90% has been eluted by 30 pore volumes. This illustrates the advantages of mobilization over solubilization. By contrast water-based dissolution would require hundreds to thousands of pore volumes; this illustrates the relative advantages of surfactant-enhanced remediation using solubilization or mobilization. The latter concentrations in Figure 6 are still several orders of magnitude above the drinking water standard. Thus, while surfactant techniques significantly expedite mass removal, extended amounts of time may be required to achieve drinking water standards (although much less than with

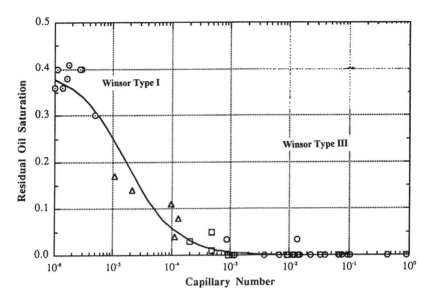

Figure 5: Residual Oil Saturation versus Capillary Number; Increasing Capillary Number Corresponds to Decreasing Interfacial Tension and Transitioning Towards Optimal Middle Phase Microemulsion (Adapted from *4*).

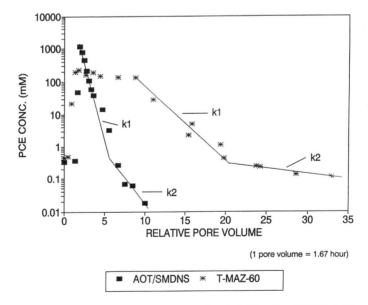

(1 pore volume = 1.67 hour)

Figure 6: PCE Elution in Column Studies for Solubilization (T-MAZ 60) and Mobilization (AOT/SMDNS) Systems

assumed that surfactant-enhanced remediation costs less than a net present cost of $5 M would offset 5 years of $1 M annual pump-and-treat costs, thereby rendering the process economical. Obviously these assumptions are limited to specific conditions, and must be adapted for a given site. It was shown that the target cost of $5 M was met for all sites of less than 1 acre in size, and for 1 acre sites with residual saturation of 5% or less. For 1/4 acre sites with residual saturations of 5% or less, the cost of solubilization remediation was only 40% of the cost of pump-and-treat. Krebs-Yuill et al. (16) also compared mobilization to solubilization. It was observed that as the site size and the residual saturation increased mobilization became increasingly attractive. For the ½ acre site at 10% residual, the mobilization net present cost was $2.1 M, while the solubilization value was $3.9 M, both below the break even point of $5 M for conventional pump-and-treat.

From these economic studies several important considerations are highlighted: minimizing surfactant losses, solubilization versus mobilization, surfactant recovery and reuse, etc. These and other items are discussed further below.

Minimizing Surfactant Losses

Minimizing physicochemical surfactant losses is a major factor in surfactant selection (e.g., precipitation, sorption, phase separation). These losses can decrease the efficiency of the system, thereby negatively impacting the economics of the system, and can even significantly reduce the conductivity of the formation, thereby rendering the process ineffective.

Nonionic surfactants typically have lower CMC values, and thus require less surfactant addition to form micelles. While nonionic surfactants do not precipitate with counterions, they are generally susceptible to higher levels of sorption and may be more subject to phase separation than anionic surfactants (17-19). Precipitation-resistant anionic surfactants may thus be ideal candidates for environmental remediation. Dowfax surfactants, with two sulfonate groups, are less susceptible to precipitation than monosulfonated surfactants (e.g., SDBS, 17). Steol surfactants have demonstrated similar precipitation resistance based on the addition of ethylene oxide groups (EO's) to the basic SDS structure (18).

Rouse et al. (17) also demonstrated that sorption of the Dowfax surfactants was less than that for SDBS (seven-fold lower), while the Steol surfactants were an order of magnitude lower in sorption than SDS (18). The sorption decreases were equally if not more dramatic when comparing the Dowfax and Steol surfactants to nonionic surfactants. In addition to the obvious economic advantages, there is also a technical advantage for reduced sorption. The sorbed surfactants, which can exist as a bilayer (admicelle) of surfactant molecules (see Figure 1), can act as an organic sink for contaminants (much as organic matter does). While this phenomenon can be exploited for sorbing barriers and separation processes (7, 20-21), its occurrence in subsurface remediation is undesirable.

Having demonstrated that the Dowfax and Steol surfactants are less susceptible to precipitation and sorption, their contaminant solubilization potential is obviously important. The Dowfax surfactants were shown to be equally if not more efficient in enhancing the solubility of contaminants relative to their monosulfonated equivalents (17). Likewise, the Steol surfactants showed comparable to slightly higher solubilization relative to their non-ethoxylated equivalents (*18*). Thus, the importance of considering surfactant losses during surfactant selection is evident.

Another concern with nonionic surfactants is phase separation. The Cloud Point is the temperature at which nonionic surfactants lose solubility and form a separate phase (*1*). Obviously, nonionic surfactants should be selected based on their Cloud Point and aquifer temperatures. Also, surfactants can form gel and numerous other phases that both deplete the surfactant from the aqueous phase, and thus diminish the desired solubilization potential, and may also adversely impact fluid mobility, etc. It is thus critical that surfactant phase behavior be carefully evaluated for site specific conditions (temperature, electrolytes, contaminants, etc.) (*22*).

In addition to physicochemical losses, hydrodynamic processes may also result in increased surfactant demand. The hydrodynamic processes that plague pump and treat operations may be amplified during surfactant enhanced remediation. For example, while injecting surfactant to target zone we desire to minimize the amount of surfactant that goes through uncontaminated media -- however, the injection-well cone of impression can work against this goal, highlighting the need for careful design and implementation. The goal will be to achieve hydraulic control (capture) while maximizing hydraulic efficiency. Hydraulic efficiency can be increased by: (1) minimizing the volume of injected surfactant solution; (2) minimizing the volume of fluid to be pumped to the surface (reducing treatment costs); (3) targeting injected chemicals to the contaminated zones of the aquifer; (4) preventing the movement of injected fluids toward clean portions of the aquifer; (5) maximizing capture of resulting water-surfactant-contaminant mixtures (*23*); and (6) use of polymers to increase sweep efficiency of the surfactant system (*4*).

Hydraulic measures that can optimize surfactant delivery include location and rates of injection and extraction wells, use of impermeable or hydraulic barriers, use of driver wells, and use of vertical circulation wells. While impermeable barriers (e.g., grout curtains, slurry walls, sheet piling) are more permanent, hydraulic controls (e.g., injection wells, infiltrations galleries) may not be totally flexible. Based on numerical simulations, Gupta and Knox (*24*) concluded that the use of dual (hydraulic and impermeable) barrier systems was most efficient in terms of reducing the volume of fresh ground water migrating through the contaminated zone and controlling migration of injected surfactant solutions. The vertical circulation well (VCW), which may be especially useful during pilot scale studies, uses simultaneous injection and extraction within a single borehole, thereby creating a circulating flow pattern. The "closed" nature of the VCW hydraulic system maximizes the capture of the surfactant and contaminant; pumping the extraction interval at a higher rate than the injection interval can further improve capture, but also draws in additional native ground water. Physical (sand tank) modeling has demonstrated advantages of VCWs relative to intrawell systems (*23*); the system has also been demonstrated at a field demonstration study in Traverse City, Michigan (*25*). In all cases mathematical modeling is critical to successful design and implementation of the system.

Subsurface flow and contaminant heterogeneities can reduce the surfactant sweep efficiency and thus require use of additional surfactant. Polymers and foam have been evaluated for improving sweep efficiency in enhanced oil recovery. By increasing the viscosity of the fluid, polymers decrease the relative permeability between pathways and serves to equalize the flow distribution through the contaminant zone. Polymers are relatively inexpensive compared to the surfactant, and the improved efficiency may more than make up for the increased costs (*4*).

Solubilization versus Mobilization

In solubilization, the more hydrophobic compounds experience the greater relative solubility enhancement (higher K_{ow} and K_m, both indicating greater contaminant partitioning into micelles). Micellar partitioning is observed to vary more between contaminants for any surfactant than between surfactants for a given contaminant (3). Thus, other factors may significantly impact surfactant selection for solubilization (e.g., cost, susceptibility to losses, toxicity, etc.). However, it should be emphasized that micellar solubilization (and thus K_m) varies as a function of contaminant type (nonpolar versus polar/ionic) and aqueous contaminant concentrations (20, 26). Aqueous contaminant concentrations below the single component water solubility may be experienced due to nonideal solubilization (i.e., nonequilibrium dissolution), mixed NAPL phases, etc. (11, 26-27).

In contrast to solubilization, surfactant(s) selection is much more dependent on contaminant composition when considering mobilization. Our initial efforts to achieve middle phase microemulsions with chlorinated solvents were unsuccessful. As the HLB of the surfactant system was varied from 2.1 to 40, a transition from a Type II to a Type I system was observed. However, a clear middle phase was not achieved in the transition region, but rather instead a liquid crystal phase occurred. When a branched surfactant, Aerosol OT (AOT), was used along with sodium mono and di methyl naphthalene sulfonate (SMDNS), a liquid crystal phase did not form and a middle phase microemulsion occurred (3). More recent experience has demonstrated that contaminant weathering can significantly impact the necessary surfactant selection to achieve a middle phase microemulsion (28). Thus, it is observed that contaminant composition and surfactant selection is critical to achieving a middle phase system.

A middle phase microemulsions was formed with 1,2-DCE using a constant AOT concentration and varying the SMDNS concentration (Figure 3). This approach was also used for forming middle phase systems with PCE, TCE, and 1,2-DCE individually (3) and in binary and ternary mixtures of these chlorocarbons (19). For mixtures, regular solution theory proved better able to predict phase behavior in the ternary system than ideal solution theory (19). It is especially encouraging that, given analyses from binary systems, we can predict results for ternary systems of variable DNAPL composition.

While design of a surfactant system is relatively robust for enhanced contaminant elution using solubilization, mobilization systems can be extremely sensitive to aquifer conditions (temperature, hardness, etc.), contaminant composition, and surfactant structure. While deviations from ideal middle phase conditions will decrease the system efficiency (much more than for solubilization), it does not necessarily negate the improved efficiency relative to solubilization. Shiau et al. (29) demonstrated that deviating just outside the ideal middle phase composition did not totally negate the enhanced removal efficiency of the system, due to the fact that interfacial tensions are still very low even if not at their minimum (see Figure 4). These results also demonstrate that solubilization and mobilization are not discrete systems, but rather are "regions" in a continuum of surfactant systems (observe the continuous decrease in interfacial tension in Figure 4). The mobilization approach also elicits concerns about vertical migration of released residual. While recognizing these

water alone). This certainly raises the issue of how clean is clean enough (i.e., risk based cleanup levels) and the potential for using coupled processes to achieve the treatment process (i.e., surfactant enhanced residual extraction followed by biopolishing, etc.).

Nonideal surfactant solubilization has been attributed to several factors, including interfacial mass transfer limitations (11), dissolution fingering and a combination of the two (27). Figure 6 provides some evidence for the nonideality by virtue of the two legs of the declining PCE elution. The dissolution fingering explanation is based upon the assumption that the NAPL distribution is heterogeneous even within apparently homogeneous systems, resulting in preferential dissolution pathways. Dissolution fingering can account for the changing length of the mass transfer zone evidenced in column studies whereas traditional mass transfer theory can not (27). Accurate scale-up and design of full scale systems will require a better understanding of nonideal dissolution and transport phenomena.

The discussions thus far have demonstrated how complex surfactant systems can be, especially mobilization. Careful design and implementation of these systems should be based on modeling efforts. These modeling efforts will likely range from one-dimensional to two- and three-dimensional efforts focusing on fluid flow and physicochemical processes. Codes such as the USGS Method of Characteristics (MOC) model may be used in addressing simple fluid flow. However, for the more complex processes associated with surfactant mobilization, a more complicated modeling approach is required. One such code is UTCHEM (The University of Texas Chemical Flooding Simulator), which is a multiphase, multicomponent, three dimensional finite difference simulator that was originally developed for surfactant enhanced oil recovery (31). In addition to hydrodynamic factors, UTCHEM also addresses interfacial tension, relative permeability, capillary pressure, capillary trapping, and phase mobilization processes. The comprehensive nature of the code makes it potentially cumbersome, but as the site complexity increases, the use of such a code becomes imperative.

Surfactant-Contaminant Separation and Surfactant Reuse

The economic analyses discussed above highlighted the importance of surfactant-contaminant separation and surfactant reuse to the economic viability of this technology, especially for multiple pore volume flushes. Two different separations may be necessary; removal of contaminant from the concentrated surfactant stream, and removal of water from the surfactant stream (concentration step).

A key to selecting the unit process for surfactant-contaminant separation is the contaminant volatility. Numerous systems can be used for volatile contaminants; however, fewer alternatives exist for nonvolatiles. Air stripping with antifoams, membrane air stripping, vacuum stripping and vacuum distillation, pervaporation, membrane liquid extraction, and precipitation are alternatives for treating volatile contaminants. The strong affinity of surfactants for activated carbon and ion exchange resins eliminates them as viable candidates for this separation. Contaminants will partition into surfactant adsorbed onto the media (admicelles), thereby preventing the desired separation (surface adsolubilization is similar to aqueous solubilization -- 20).

For high volatility contaminants conventional packed-column air stripping can be used. Lipe et al. (*32*) showed that air stripper efficiency declined with increasing surfactant concentration, and that the impact was more dramatic for more hydrophobic contaminants and for surfactants with increased solubilization potential. The reduced contaminant activity, which results from micellar solubilization, is responsible for these observations. Predictions using air stripper design equations modified to account for micellar solubilization agreed well with experimental results. Using the modified design equations will allow proper design of air strippers to achieve the desired effluent concentrations, taking into account surfactant impacts on the system.

Foaming is a problem that may be encountered in packed-tower air strippers. The air to water ratio can be decreased in an attempt to avoid heavy foams in the column (*32*). However, lowering the air to water ratio will decrease the efficiency of the system, thereby increasing the required tower height. Antifoams can be used to mitigate the problem without lowering the air flow rates, but implications of using an antifoam must be considered (i.e., reinjection and discharge permits). Vacuum stripping and vacuum distillation may extend the range of contaminants that can be removed, but again foaming may be an issue. The use of hollow fiber columns can mitigate the foaming problem (*32*). without sacrificing efficiency and without requiring the use of antifoams. It may be necessary to control the pressure difference across the porous hollow fibers to prevent weeping of the surfactant solution across the membrane.

Liquid-liquid extraction can be used for non-volatile contaminants. Hollow fiber membranes will eliminate concerns of emulsion formation in this process. Hasegawa et al. (*33*) showed that the hollow fiber liquid-liquid extraction efficiency decreased with increasing surfactant concentration, for more hydrophobic contaminants, and for surfactants with increased solubilization potential. As in the case of air stripping above, these results are expected due to micellar solubilization. By incorporating micellar solubilization into the liquid-liquid extraction design equations the model predictions agreed well with experimental results. Using the modified design equations will allow proper design of the liquid-liquid extraction process to meet treatment goals. Choice of the extracting oil will be based upon minimal water solubility, environmental acceptance and efficiency in extracting the contaminant.

For cases where surfactant reconcentration is necessary, ultrafiltration (UF) can be used. The UF process will retain surfactant micelles while allowing monomers to pass through (e.g., a filter with a 10,000 molecular weight cutoff -- MWCO). Lipe et al. (*32*) achieved 93 to 99% retention of surfactants using a 10,000 MWCO membrane. While the UF process effectively retains surfactant micelles, monomeric concentrations will pass through with the permeate. If economics dictate, smaller filters may be used to capture this additional surfactant for reuse (*15*).

Given that surfactant reuse is crucial to the economic viability of the system, contaminant-surfactant separation and surfactant reuse needs to be a key consideration in the surfactant selection process. The ultimate goal is a surfactant system that is highly efficient in extracting the contaminant, has favorable phase behavior in the subsurface, and can efficiently be decontaminated and reused. Failure in any of these areas can render the system uneconomical and/or infeasible.

Environmental Acceptability

The environmental acceptability of surfactants is critical to their widespread use. Obvious issues regarding surfactant injection are the short term and long term risks associated with the surfactant. A main surfactant toxicity concern is their impact on aquatic life; they may be toxic to fish at the ppm levels. As such, if ground water-surface water interactions exist at or near the site, this needs to be given careful consideration. Relative to human health, surfactants are not typically regulated and as such health based standards do not exist. This concern might suggest the use of surfactants with US Food and Drug Administration direct food additive status (*3, 19, 22*). Not only have these surfactants been approved for direct consumption, the fact that many of them are combinations of fatty acids and sugars alleviates concerns as to their degradation products. However, caution must be exercised when considering the value of this designation. For example, surfactants have been developed that have improved characteristics (e.g., improved resistance to sorption and precipitation) but that have not obtained, or even pursued, direct food additive status. Manufacturers are hesitant to invest the financial resources necessary to obtain direct food additive status; as such, numerous surfactants may qualify for this status, but have not undergone the necessary testing. Surfactants with indirect food additive status have been approved for contacting food products (i.e. for washing the product or use in packaging); a number of these have better operating characteristics versus surfactants with direct food additive status (*19*). Obviously, as this technology progresses, improved methods for assessing environmental friendliness are needed.

Surfactant degradation raises several issues; what are the rates under various redox conditions, what are the metabolites, and what are the concerns associated with the answers to these questions. While a significant amount of data exists on the degradation of surfactants in aqueous aerobic systems, much less data exists on the degradation of surfactants in soil environments under varying redox conditions. A number of ongoing efforts are evaluating these issues; however, additional research is needed. The ease of surfactant degradation requires a balance; the surfactant should not be so degradable that its prematurely lost or that it causes bioplugging, yet it should be sufficiently biodegradable so as not to persist after the remediation is completed. The higher concentrations during the remediation may limit biodegradability, whereas the lower post-remediation concentrations may be more amenable to biodegradation. It may be possible that, if properly designed, biodegradation of the residual surfactant might help to promote biopolishing of residual contaminant.

Summary

This chapter has provided an overview of surfactant fundamentals, and highlighted issues important to the selection and design of surfactant systems. Careful consideration of surfactant, contaminant, media and fluid properties are critical to the design of a successful, and optimal, surfactant system. These properties should be evaluated not only individually, but collectively. By integrating the design approach in this manner, using laboratory batch, column and modeling studies, surfactant systems can be developed that will significantly improve the remediation system. Field demonstrations of this approach are the subject of subsequent chapters.

Acknowledgments

We would like to thank Drs. Shiau, Rouse, and Soerens, and Mr. Wu, along with a other graduate and undergraduate researchers for their faithful and diligent research efforts. We would also like to acknowledge research funding from numerous federal (e.g.,

SERDP, EPA, DOD, DOE) and industrial sources (including industrial sponsors of the Institute for Applied Surfactant Research). The lead author would also like to thank Dr. Scott Huling (Kerr Environmental Research Center) for his insightful review comments.

References

1. Rosen, M. J. *Surfactants and Interfacial Phenomena.* 1989, 2nd ed., John Wiley and Sons, Inc., New York, NY.

2. Edwards, D. A., Luthy, R. G. and Liu, Z. *Environmental Science and Technology.* 25(1), 1991, 127-133.

3. Shiau, B. J., Sabatini, D. A., and Harwell, J. H. *Ground Water.* 32(4), 1994, 561-569.

4. Pope, G. and Wade, W. in *Surfactant-Enhanced Subsurface Remediation: Emerging Technologies.* Sabatini, D., Knox, R. and Harwell, J. eds., ACS Symposium Series 594, American Chemical Society, Washington, DC, 1995, 142-160.

5. Bourrel, M. and Schechter, R. S. *Microemulsions and Related Systems.* Surfactant Science Series, Vol. 30, Marcel Dekker, Inc., New York, 1988.

6. Valsaraj, K. T., and Thibodeaux, L .J. *Water Research.* 23(2), 1989, 183-189.

7. Smith, J. A. and Jaffe, P. R. *Environmental Science and Technology.* 25, 1991, 2054-2058.

8. Jafvert, C. T. and Heath, J. K. *Environmental Science and Technology.* 25(6), 1991, 1031-1038.

9. West, C. C. and Harwell, J. H. *Environmental Science and Technology.* 26, 1992, 2324-2330.

10. Abdul, A. S., Gibson, T. L., Ang, C. C., Smith, J. C. and Sobczynski, R. E. *Ground Water,* 30(2), 1992, 219-231.

11. Pennel, K. D., Abriola, L. M. and Weber, W. J. Jr. *Environmental Science and Technology.* 27(12), 1993, 2341-2351.

12. Baran, J. R., Pope, G. A., Wade, W. H., Weerasooriyaa, V. and Yapa, A. *Environmental Science and Technology.* 28(7), 1994, 1361-1366.

13. Fountain, J., Waddell-Sheets, C., Lagowski, A., Taylor, C., Frazier, D. and Byrne, M. in *Surfactant-Enhanced Subsurface Remediation: Emerging Technologies.* Sabatini, D. A., Knox, R. C. and Harwell, J. H., eds., ACS Symposium Series 594, American Chemical Society, Washington, DC, 1995, 177-190.

14. Sabatini, D. A., Knox, R. C., and Harwell, J. H., eds. *Surfactant Enhanced Subsurface Remediation: Emerging Technologies.* ACS Symposium Series 594, American Chemical Society, Washington, D.C., 1995.

15. Krebs-Yuill, B., J. Harwell, D. Sabatini, and R.Knox. in *Surfactant-Enhanced Subsurface Remediation: Emerging Technologies,* D. Sabatini, R. Knox, and J. Harwell (editors), American Chemical Society, Washington, D. C., 1995.

16. Krebs-Yuill, B., J. Harwell, D. Sabatini, G. Quinton, and S. Shoemaker, "Economic Study of Surfactant-Enhanced Pump-and-Treat Remediation," 69th Annual Water Envir. Federation Conference, Dallas, Texas, October 5-9, 1996.

17. Rouse, J. D., Sabatini, D. A. and Harwell, J. H. *Environmental Science and Technology.* 27(10), 1993, 2072-2078.

18. Rouse, J. D., Sabatini, D. A., Brown, R. E. and Harwell, J. H. *Water Environment Research*, 68(2), 1996, 162-168.

19. Shiau, B. J., Sabatini, D. A., Harwell, J. H. and Vu, D. *Environmental Science and Technology*, 30(1), 1996, 97-103.

20. Nayyar, S. P., Sabatini, D. A. and Harwell, J. H. *Environmental Science and Technology*. 28, 1994,1874-1881.

21. Bowman, R. S., Haggerty, G. M., Huddleston, D. N. And Flynn, M. M. in *Surfactant-Enhanced Subsurface Remediation: Emerging Technologies.* Sabatini, D. A., Knox, R. C. and Harwell, J. H., eds., ACS Symposium Series 594, American Chemical Society, Washington, DC, 1995, 54-64

22. Shiau, B. J., Sabatini, D. A. and Harwell, J. H. *Environmental Science and Technology* 29(12), 1995, 2929-2935.

23. Chen, Y., Chen, L. Y and Knox, R. C. in *Surfactant-Enhanced Subsurface Remediation: Emerging Technologies.* Sabatini, D. A., Knox, R. C. and Harwell, J. H., eds., ACS Symposium Series 594, American Chemical Society, Washington, DC, 1995, 249-264.

24. Gupta, H. and Knox, R. C. "Modeling the Effectiveness of Hydraulic and Impermeable Barriers for Improving Contaminant Extraction or Surfactant Injection," PETRO-SAFE 96, ASME Petroleum Divising and American Petroleum Institute, Houston, TX, 1996.

25. Knox, R. C., Sabatini, D. A., Harwell, J. H., Brown, R. E., West, C. C., Blaha, F., and Griffin, S. "Surfactant Remediation Field Demonstration Using a Vertical Circulation Well." In Press, *Ground Water*, 1997.

26. Rouse, J. D., Sabatini, D. A., Deeds, N. E., Brown, E., and Harwell, J. H. *Environmental Science and Technology* 29(10), 1995, 2484-2489.

27. Soerens, T. S. "Dissolution of Nonaqueous Phase Liquids in Porous Media: Mass Transfer in Aqeuous and Surfactant Systems." PhD Dissertation, 1995. University of Oklahoma, Norman, OK.

28. Wu, B., Shiau, B. J., Sabatini, D. A., Vu, D. Q. and Harwell, J. H. "Formulating Microemulsion Systems of Petroleum Hydrocarbons using Surfactant-Cosurfactant Mixtures." In Review, *Ground Water*, 1997.

29. Shiau, B. J., Sabatini, D. A. and Harwell, J. H. "Removal of Chlorinated Solvents in Subsurface Media Using Edible Surfactants: Column Studies." Accepted *Journal of Environmental Engineering Division - ASCE*, 1997.

30. Morrow, N. R., Chatzis, I. and Taber, J. J. *SPE Reservoir Engineering*, 3(3), August 1988, 927-934.

31. Delshad, M., Pope, G. A., and Sepehirnoori, K. A. *J. Contaminant Hydrology.* 23, 1996, 303-327.

32. Lipe, M., Sabatini, D. A., Hasegawa, M., and Harwell, J. H. *Ground Water Monitoring and Remediation*, 16(1), Winter 1996, 85-92.

33. Hasegawa, M., Sabatini, D. A. and Harwell, J. H. "Liquid-Liquid Extraction for Surfactant-Contaminant Separation." In Press, *Journal of Environmental Engineering Division - ASCE*, 1997.

Chapter 3

Enhanced Recovery of Organics Using Direct Energy Techniques

T. R. Jarosch and B. B. Looney

Department of Environmental Science and Technology, Savannah River Technology Center, Westinghouse Savannah River Company, Aiken, SC 29808

Enhanced recovery of residual non-aqueous phase liquids (NAPL) can be achieved by directing energy into the contaminated subsurface. Proposed methods include low frequency electrical energy or high frequency electromagnetic energy to heat the sediments (joule or resistive and dielectric heating), an induced dc voltage gradient to promote water or ion movement (electro-osmosis and -migration), or vibrational and sonic energy to enhance contaminant release and convection. Each of these technologies offer potential enhancements to conventional extraction methods - soil vapor extraction (SVE) above the water table, and groundwater pump and treat below the water table. The general goal of direct energy methods is to overcome mass transfer limitations and other geological, physical, and chemical limitations. A brief description of each technology, its applicability (including strengths, weaknesses, and limitations), operational requirements and status, as well as areas of on-going or potential future research are presented.

As with many environmental remediation technologies for non-aqueous phase liquids (NAPL), direct energy applications are adaptations of techniques originally developed in the oil field or construction industry (1-3). The technologies presented here, soil heating, electro-osmosis, and sonic (acoustic) enhancement, are at varying stages of development ranging from commercially available to bench scale. The discussion is focused on describing the basic physical concepts of each technology, providing general information on the applicability and limitations of each technology, and listing some of the testing which has occurred.

Soil Heating

The soil heating technologies involving direct energy injection are most applicable in enhancing soil vapor extraction (SVE) operations. There are a number of phenomena and issues related to soil heating for enhanced vapor removal which are independent of the particular heating mechanism (direct energy or hot fluid). Conventional SVE of organics is enhanced through thermal processes by a number of phenomena, including increased vapor pressure, increased solubility, increased desorption (from sediments), increases in the effective vapor permeability through the removal (by steam production) of pore water, steam displacement, and steam stripping. How much energy is required is a function of several factors including the soil composition (which, in turn, determines its heat capacity), the desired average soil temperature, and the inefficiencies and/or heat losses involved with a particular thermal application or plume site. An important factor in implementing soil heating is the relatively high heat capacity of water (roughly 3 to 4 times that of most sediments) and the high latent heat of vaporization of water assuming water removal or steam production is desired. For instance, to raise the temperature of a nearly saturated (85%) sandy clay from 20 $^{\circ}$C to 100 $^{\circ}$C and vaporize all the pore water requires an energy input, disregarding losses, of 180 to 200 kWH per cubic yard (a cu. yd. under such conditions is approximately 1.4 metric tons). While there are obvious benefits to be realized by vaporizing pore water, VOCs have boiling points typically below 125 $^{\circ}$C, and a significant enhancement to remediation can be accomplished without reaching this temperature. Furthermore, by raising subsurface temperatures to 100 $^{\circ}$C, the vapor pressures of semivolatile organic compounds (SVOCs) will be similar to VOCs at ambient temperature *(1)*. In addition to simple thermodynamic concerns, thermal enhancement will also tax conventional SVE systems in a number of critical areas. First, an additional condenser must be added to the recovery train to accommodate the steam extracted. Obviously, surface equipment to store and/or treat the water recovered is also necessary. Second, the choice of off-gas treatment must be robust enough to handle higher and often highly varying concentrations. Third, the vacuum extraction pump(s) must also be robust enough to adequately handle the higher vapor flow required when or if steam production starts. Consideration must also be given to induced fluid movement during thermal applications (e.g. rising hot vapors, thermal convection in groundwater, etc.) which must be taken into account when deciding where to locate extraction wells and screens *(4)*.

Radio Frequency Heating. Radio frequency heating is an electromagnetic (EM) technique similar to microwave heating. The heating occurs internally through a dielectric mechanism in which the molecular dipoles interact with the EM wave. The resulting molecular distortion or motion induced by the interaction is translated from mechanical to thermal energy. The effectiveness of the dipole coupling and the power absorbed is a function of the frequency and amplitude of the electric field and the dielectric properties of the sediments which, in turn, are a function of the soil composition, temperature, and EM frequency. The absorption efficiency is directly related to the imaginary part of the dielectric constant or the loss tangent (defined as the ratio of the imaginary to the real part of the dielectric constant) of the propagating media. While certain sediments and even bedrock will absorb radio waves to some extent, the bulk of the absorption in the soil occurs from pore water or physically adsorbed water. Water, being a very polar molecule, has a high loss tangent. Despite the reliance on water for energy absorption, dielectric heating does not require a continuous conductive path of pore water, and the soil matrix continues to absorb sufficient EM energy to attain temperatures ranging from 300 to 400 °C even after most of the water has boiled away. The most applicable transmission frequencies for soil remediation applications are in the 1 to 100 MHz range. Higher frequencies are more readily absorbed by the sediment matrix, but there is a trade off in the radial influence. The radial influence, often defined as the skin depth (where the field strength has been attenuated by a factor of 1/e) is inversely proportional to the square root of the frequency. However, the effective radial influence of the EM field will grow as the soil near the electrode or applicator dries and becomes less lossy (lower loss tangent).

There are two common methods for delivering radio frequency power to the soil, via a dipole applicator(s) or via a triplate excitor array(s). The dipole applicator is a collinear loaded antenna fed by a coaxial cable. The diameter and length of the applicator is determined by the frequency and dielectric properties of the soil (higher frequencies typically require shorter applicators with larger diameters). The applicators can be deployed in conventional wells provided they are completed with casing materials compatible with the RF field (e.g. non-conducting) and are capable of withstanding high temperatures. The lengths of a single applicator can be adjusted in the field and applicators can be changed out as the soil properties change during heating. Applicators can be used in an array or with reflector electrodes to focus or steer the energy. The applicator spacing is determined by the frequency and dielectric properties but generally ranges from 5 to 15 feet. The triplate design is the rectangular analog of a cylindrical coaxial cable in which a bound wave is established between a central "planar" conductor consisting of a row of vertical excitor electrodes and two outer rows acting as ground electrodes or wave guides. The electrodes are constructed of simple copper or aluminum tubing and set to depths of up to 30 feet. The spacing between excitor electrodes and their corresponding ground electrodes is determined by the applied frequency and soil dielectric properties and, as with the dipole applicator, ranges from 5 to 15 feet.

Several pilot-scale remediation tests have been conducted in the field with both application techniques. The triplate system developed by researchers at the Illinois Institute of Technology Research Institute (IITRI, Chicago, IL) has been tested at three DoD sites [at Volk AFB (5), Rocky Mountain Arsenal, and one in cooperation with the EPA SITE Program at Kelly AFB (6)]. ITTRI has also recently completed a test with Sandia National Laboratory (DOE) of the Thermal Enhanced Vapor Extraction System (TEVES) which combined (consecutively) resistive and RF heating (7). General Electric Corp. is investigating marketing the triplate system with solid state RF generators (8). The dipole applicator system designed by researchers at KAI Technologies, Inc. (Portsmouth, NH) has been tested at SRS (DOE) in a horizontal well application (9) and at Kelly AFB (also with the EPA Site program) (10).

Radio frequency heating is most applicable for vadose zone remediation, preferably in soils with low water content (ideally < 10 % by volume). For soils with high water content or in the saturated zone, remediation could be accomplished more quickly with dewatering (if feasible) or by first using resistive (joule) heating to drive off the bulk of the water assuming the maximum temperature goal is in excess of 100 °C. The technology is effective in the remediation of VOCs, SVOCs, and high boiling compounds in low or high permeable sediments. The dipole technique is applicable at any depth and the applicator can be moved within a single well to heat a larger volume. For deep applications, steam condensation in the vapor extraction well may entail the installation of adequate sumps with down-hole pumps. The dipole technique is compatible with conventional and/or directional drilling methods that make it applicable for remediation of plumes in poorly accessed areas (e.g. under buildings). The triplate excitor array technique is only applicable for shallow plumes (< 50 feet). However, this method appears to be more compatible with larger powered RF systems (e.g. > 50 kW). Based on field tests, dipole systems may be limited to powers ≤ 25 kW due to overheating of the applicator.

Surface equipment includes an RF generator, matching network, dummy load, controls, and diagnostics all of which can typically be housed in a small trailer on the order of 8' x 8' x 20' (assuming generator powers less than 50 kW). Beyond this equipment, the dipole applicator needs little else besides space for running transmission lines. The triplate technique will typically require good surface access and additional drilling. The spacing between plates are roughly equivalent to spacing between applicators, however, the spacing of electrodes within a row is half the spacing between plates, therefore the triplate design would require approximately twice the number of boreholes used in an applicator array design. In both applications, while electrode or applicator wells can serve as vapor extraction or air sweep sources, additional wells should be installed to ensure adequate extraction and vapor containment. In shallow applications, a surface barrier over the heated zone will be required to act as a thermal insulator, as a shield for stray radio wave emissions, as a vapor recovery plenum, and to prevent water infiltration.

The major limitations of this technology are threefold. First, the thermal radial influence per dipole or electrode is relatively small. Therefore, the techniques require considerable drilling to place wells and electrodes which is often a major cost in remediation economic analyses. Second, large RF generators required to handle typical remediation soil volumes are either not reliable or very expensive (total costs for generators and matching networks range from $3 to $4 per watt, roughly an order of magnitude greater than line power heating equipment). Third, dielectric heating has a relatively poor energy efficiency with respect to AC power (overall efficiencies are typically < 70%). This factor, combined with the high capital cost make it impractical for removing (boiling off) large volumes of water, or remediating large plumes (> 1000 tons or cubic yards). Minor limitations include hazards associated with stray surface emissions in shallow applications or of coupling with buried conductors that can carry the field to the surface or result in local hot spots within the targeted heated zone. However, these hazards can be mitigated and are not as acute as those associated with power line resistive heating.

Dielectric heating is one of the most practical techniques of attaining soil temperatures well in excess of 100 °C in low permeable soils. The heating occurs internally and absorption will generally be focused in the lower permeable soils with higher water saturation. These low permeable zones also tend to retain high levels of residual organics with concentrations in excess of what can be explained simply from water solubilities. The heating mechanism does not require a continuous conductive path (e.g. no water injection is required). The dipole applicator system provides a means of focusing or steering the energy and is compatible with directional drilling methods to access remote plumes (e.g. under buildings or tanks).

Further development in this area is needed on low cost, high power, solid state RF generators *(8),* testing of automated matching networks for high power, high dynamic range applications, and field testing of multiple phased applicator arrays. Development of models which combine fluid, thermal and contaminant flow to ensure proper capture and/or containment is a universal need for all heating technologies *(4).* While not a significant factor in most remedial applications, research is needed on the physical and chemical characteristics of the subsurface following heating at high temperatures (> 250 $^{\circ}$C). Investigations of alternative well completion materials for the dipole technique could alleviate the adverse effects of high temperatures achieved at the applicator.

Microwave Heating.   From a physical phenomena standpoint, the microwave heating process is essentially the same as heating with radio waves, simply conducted at a higher frequency (on the order of GHz).   At these frequencies, there is strong absorbance contribution affecting molecular orientation (RF fields distort dipoles), and consequently, microwaves are more readily or effectively absorbed by water. This factor has previously retarded the use of microwaves for bulk heating because absorption was so strong that the effective skin depth or field penetration was too small to heat a large volume. The current line of approach to overcome this effect is, essentially, the use of high power generators (continuous wave klystron tubes) with outputs on the order of 0.5 to 2 MW (Peter Kearl, Oak Ridge National Laboratory, Grand Junction CO, personal communication, 1996). These types of power sources have been developed in the defense industry of Russia for military radar sites. The system would operate with a down-hole antenna similar to the RF dipole applicator technique. However, the well construction at the applicator, as currently proposed, would consist of a fused quartz casing which possesses the required dielectric and thermal properties to match the impedance and power of the antenna.   Current research claims that the higher power sources will induce and sustain a growing phase boundary at the outer extent of the induced field which consists of hot vapor under several atmospheres of pressure. The dry soil within this phase boundary is relatively transparent to the microwave EM field. Vapors will move from the high pressure boundary to the well for recovery. The effective radial influence of a single antenna is estimated to range from 6 to 30 meters for 0.5 MW to 2 MW generators, respectively.

The technology, which is still in the R&D stage, is being developed by researchers at Oak Ridge National Laboratory (Grand Junction, CO) and West Coast Applied Physics (Santa Rosa, CA). At present, research work has consisted of bench scale treatability studies of soil cores using a microwave cavity and modeling. The current system, as proposed, is applicable in both the vadose and saturated zones and is independent of soil permeability.

Equipment similar to that employed with RF heating is required. However, due to the increased power levels, the foot print and service/utility requirements will increase substantially. As with RF heating, this technology is anticipated to require substantial power consumption with poor efficiency (based on wall power to soil energy conversion). In addition, full development will entail significant capital and development costs to mass produce klystron tubes which, at present, have a suspect reliability and limited lifetime.

Resistive Heating. Resistive or joule heating is an internal heating technique in which an AC current is passed through the soil and energy is dissipated through ohmic losses. The method employs normal line power (60 Hz) applied to electrodes placed in the ground. The power input to the soil is inversely proportional to the soil resistivity and directly proportional to the square of the applied voltage. The resistivity of the sediment matrix is largely determined by its water content with an obvious contribution from dissolved salts or ionic content and additional effects brought on by the ion exchange capacity of the sediments themselves. In addition, the resistivity is a function of temperature, and as the water heats up to its boiling point, the conductivity increases with increased ion mobility. The applied voltage is limited by the total resistance, a function of the resistivity of the soil and geometry of the electrode system. For power inputs on the order of tens or hundreds of kilowatts, applied voltages range from 100 to 1500 V with total resistances of ten to a few hundred ohms. Because the current density is highest at the electrodes, the applied voltage is critically dependent on the contact resistance. To prevent excessive drying or voltage breakdown at the electrodes, water is typically injected to maintain good electrical contact. This may be augmented with low concentrations of salt added to the water and/or the use of highly conductive packing (e.g. carbon/graphite) around the electrodes. The simplest design uses single phase power fed to a single or row of vertical conductor electrode(s) with a corresponding ground receptor electrode(s). The spacing between electrodes is obviously dependent on the soil resistivity. However, the effectiveness of the method in uniformly heating the soil is dependent on a homogeneous current path, a feature seldom realized in the field due to the heterogeneity of the sediments. To improve the uniformity of heating and reduce local current densities at the electrodes (i.e. increase available current pathways), more recent designs employ multiple phased arrays of electrodes with a central ground electrode that typically doubles as a vapor extraction well. Three phase systems are the easiest and most direct power sources. Additional heating uniformity can be obtained by alternating between nested three phase systems laid out in an hexagonal array or by adding transformers to generate true six-phase power. With such techniques, single arrays up to several tens of feet in radius are feasible.

The technology is at the field deployable stage. Both three phase and six phase heating have been tested in relatively large scale applications at DOE sites. Three phase heating developed by researchers at Lawrence Livermore National Laboratory (Livermore, CA), was successfully tested at that site in both an engineering scale test targeting TCE in the vadose zone and in a full-scale test in conjunction with steam injection targeting an LNAPL plume that extended into the water table *(11)*. Full-scale six phase soil heating (SPSH), developed by researchers at Battelle Pacific Northwest

National Laboratory (PNNL, Richland, WA), was successfully tested at the Savannah River Site targeting PCE and TCE in a 10 foot thick clayey region approximately 35 feet below ground surface *(12)*. Battelle PNNL has formed a teaming agreement with Terra Vac, Inc. (San Juan, PR) to market the SPSH technology commercially.

As with dielectric heating, the resistive technique is most applicable in the vadose zone. However, because of the need for a continuous conductive path, the method is more applicable in low permeable sediments which typically contain (and will maintain) high levels of saturation. These same zones typically act as major sources for high levels of residual organics. Because its overall energy efficiency is greater than with dielectric heating, the technique can be applicable in the saturated zone, but it is probably most effective with light non-aqueous phase liquids (LNAPLs) in applications where vapor extraction is the primary recovery method. There is no depth limit, however, as with all heating methods, modifications to extraction wells must be made in deep applications to handle steam condensate produced en route to the surface.

Surface equipment varies depending on the specific method, site, and scale. The technology directly employs utility (60 Hz) electrical power or requires relatively simple and easily maintained power conversion equipment. As a result, the technique is more amenable to high powered applications (> 100 kW) often required to heat large volumes of soil. For single or three phase heating, little more than power feed lines, voltage regulating or variegating devices, and/or a generator is required. Depending on soil properties, single arrays up to 100 feet in diameter can be operated. Six phase heating will require additional space for a transformer which can also be designed to include voltage controls. All large scale (> 100 kW) resistive heating methods benefit if 13.8 kV or higher power line connections are near the sight. The plume must be accessible from the surface to accommodate electrode installation. Furthermore, due to safety concerns with regard to high voltage potentials in surface work areas and/or the ability for buried conductors to carry high voltage potentials out of the immediate remediation area, extreme care must be taken in applying the technology in heavily developed or industrial areas. Additional equipment is required for water (or brine) injection at the electrodes.

A limitation to the technology's applicability in permeable sediments is low saturation levels and the inability to control excessive drying or voltage breakdown without injection of large volumes of water. However, VOCs in high permeable sediments are adequately remediated with SVE alone and these conditions are also more amenable to injection of hot fluids (e.g. steam). The requirement for water or brine injection will increase waste water generation and may raise concerns with regulators. The method is limited to remediation temperatures at or slightly above 100°C. However, as mentioned previously, the recovery of co-contaminant SVOCs are still likely to be greatly enhanced at these temperatures.

The primary research areas for this technology include simplified electrode design and installation, field testing with dual or multiple phased arrays, field testing below the water table in conjunction with dewatering and or groundwater recovery, and testing on the effectiveness with SVOCs and high boiler co-contaminants.

Electro-osmosis

Electro-osmosis is an electro-kinetic phenomena in which ions in the diffuse double layer near soil particles move in response to an applied dc field and water movement in the soil pores is induced in a parallel direction via shear forces or viscous drag at the double layer interface. Typical soils carry a negative surface charge, so that the double layer consists of positively charged ions which will move toward the negative electrode (cathode). The electro-osmotic conductivity is defined analogously with hydraulic or electrical conductivity, a coefficient that relates flow or flux to the driving force, in this case, the voltage gradient. Under conditions typically found in saturated sediments, the electro-osmotic conductivity is directly proportional to the permitivity of the solution (a function of the real part of the dielectric constant), and the zeta potential of the soil (the potential difference between the surface potential and the potential at the shear plane), and inversely proportional to the viscosity of the fluid. While the conductivity is also related to the soil porosity, it is essentially independent of pore size or the hydraulic conductivity for low permeable sediments (hydraulic conductivity $< 10^{-3}$ cm/s). For typical sediments with hydraulic conductivities ranging from 10^{-9} to 10^{-4} cm/s, electro-osmotic conductivities range from 1 to 10 x 10^{-5} cm^2/volt/second. Therefore, for reasonable induced voltage gradients on the order of a V/cm, electro-osmotic velocities can equal or even exceed hydraulic velocities by several orders of magnitude in low permeable sediments (assuming reasonable induced hydraulic gradients of 0.01 cm/cm). While these velocities are relatively independent of the soil heterogeneity, they amount to no more than a couple inches per day. In addition to electro-osmotic flow, however, water electrolysis reactions at the electrodes affect large changes in pH. This can result in unstable operation with metal dissolution in low pH conditions generated at the anode or metal precipitation in high pH conditions generated at the cathode. These drawbacks can be circumvented by several means including reinjecting water removed at the cathode to the anode, direct conditioning with acidic or basic solution injection at the appropriate electrodes, or periodically reversing the polarity of the system. For organics removal, the method is limited to the soluble fraction of NAPL. However, the method can be combined with cosolvent or surfactant injection (the latter would also involve electrophoretic mobility) to aide in organic recovery. In addition to induced electro-osmotic flow, the method also induces electromigration of ions, which can be beneficial if metal co-contaminants need to be recovered. In its simplest application, water driven to the cathode would be pumped to the surface for treatment. The process can be combined with recovery (e.g. adsorbent wicks) or in-situ treatment walls (metal catalysts, oxidants, etc.) sandwiched between electrodes. This layered technique has been aptly labeled the Lasagna process. The method may allow the electrodes to be placed outside the primary plume area, is compatible with polarity reversal to overcome pH gradient problems which, in turn, permits multiple passes through the treatment/recovery zone(s).

The basics of the technology are well established from industrial applications in dewatering and clay consolidation (2). Applications for heavy metals removal are available commercially through a European firm. The Lasagna process for remedial applications is being developed by a consortium of industry researchers with

Monsanto, General Electric, and E.I. DuPont de Nemours. An engineering-scale test of a Lasagna system employing steel electrodes and GAC wick drain treatment zones was tested in early 1996 at a site contaminated with TCE in the Paducah Gaseous Diffusion Plant. The test was successful in removing more than 98% of the TCE with only three pore volumes of water flushed through the cell (J.W. Douthitt, Lockheed-Martin Energy Systems, Oak Ridge, TN, personal communication, 1996). A second phase of testing is planned with a much larger cell and zero valent iron catalyst treatment zones.

The technology is most applicable for remediation in the saturated zone with low permeability (hydraulic conductivities $< 10^{-3}$ cm/s). Within these bounds, the induced flow is independent of soil heterogeneity. The method has a low power consumption, will induce a uniform flow independent of soil heterogeneity, and can include or enhance in-situ treatment of NAPLs. Because it enhances extraction of the dissolved phase only, it has limited applicability in plumes with large contributions of NAPLs in pools or trapped in isolated pores unless combined with other enhanced recovery techniques such as cosolvent or surfactant addition. It can be used in plumes containing both organics and metals (including radionuclides). It may be applicable in a complimentary mode with other methods that would benefit from dewatering in the vadose zone. The size of any given cell is dependent on the soil electrical and hydraulic properties and limited by reasonable electrode voltages (100's of volts) and gradients (< 1 V/cm) to dimensions on the order of tens of feet in the direction parallel to flow. Therefore, for large plumes, the method may entail considerable drilling and excavation to install multiple cells.

The major requirement for the method is sufficient space for installing electrode and treatment/recovery walls. Any of the developed or developing barrier wall installation techniques are compatible with the technology. The choice is dependent on the size of the site, the soil properties, and the design of the electrode or treatment/recovery zone. Electrical surface equipment includes rectifiers and voltage controls. Unless in-situ treatment walls are employed, equipment to treat the water at the surface is required. In addition, contingencies may be required to remove, replace, and/or treat materials used in recovery/treatment zones.

The method has low water removal rates, however, in the types of sediments it is applicable in, the induced flow can be significantly larger than achievable with hydraulic methods. There can be significant logistics and costs involved with electrode and treatment zone installation. It can induce adverse chemical changes that may require injection of new solutions to the subsurface.

This technology is less mature than most of the soil heating techniques and, as such, development still entails continued field and scale-up testing. Investigations are focused on alternative treatment and/or recovery zones, electrode design, pH gradient control techniques, testing with metal co-contaminant removal, and testing with co-solvent or surfactant additions.

Sonic (Acoustic) Enhanced Remediation

Several organizations are investigating the potential for enhancing NAPL removal from the vadose zone and groundwater using directed acoustic energy. These organizations include Weiss and Associates (Emeryville, CA), the University of Delft

(14), and Battelle Memorial Institute *(15)*. Previous studies in oil field recovery are also relevant *(3)*. Three types of enhancement mechanisms are proposed. First, increases in mass transfer at the interfacial pore throats results from physical vibration of the pore fluid are possible. Second, improved flow properties (due to vibrational alignment, disintegration of pore blocking aggregate, cavitation, and temporary agitation) have been suggested. Third kinetic energy interactions with the solid matrix resulting in increases in temperature and pore pressure are cited as a possible mechanism for decreased DNAPL viscosity, mobilization of "sorbed" contaminants, and remobilization of trapped residual saturation DNAPL. Kim et al. *(15)* propose using acoustic energy in combination with electro-osmosis for improved remediation performance. The remaining investigators propose direct acoustic enhancement of either SVE or groundwater pump and treat. In these technologies the enhancing agent is sound waves that are delivered using an applicator as discussed below. Most of the proposed applications can be considered to be conceptual in nature and the proposed mechanisms have not been confirmed.

With the exception of oil field testing and pilot testing in the Netherlands *(14)*, there has been limited operation of acoustic enhancement. Oilfield tests showed measurable (but modest) improvements in recovery and the technique was not widely applied because alternative technologies (steam, surfactants, cosolvents, etc.) were more effective. As typically described, acoustic energy is directed into the contaminated formation using an applicator, often a cross over tool from borehole seismic work. Example tools include compressed air and hydraulic pistons/orbiters, electromechanical solenoids and orbiters, piezoelectric transducers, combustion pulses, and some additional experimental methods. Costs for the energy range from 0.1 to 100 $/kW with typical values ranging from 1 to 10 $/kW. Phased arrays are possible with some of the sources to focus the energy on selected locations. There is little data available on power needed and expected performance to facilitate implementation. Implementation of acoustic augmentation of electrokinetic and electro-osmotic processes has currently been documented only in general terms.

Acoustically enhanced remediation has been proposed for LNAPL and DNAPL above and below the water table in unconsolidated sediments. The technology is not applicable to fractured rock. Limited laboratory and field data for LNAPLs suggests limited effectiveness with variable recoveries ranging from 0 to 25%. The enhancement occurs early as the energy impacts NAPL trapped in the large pores. The NAPL that is not influenced relatively quickly may be resistant to additional acoustic stimulation. One could assume a similar performance for DNAPL although there is insufficient data. The acoustic enhancement would not alter the need to continue pump and treat or implement a more effective DNAPL recovery. Despite early claims of the developers, there is no data from the laboratory or field for DNAPL recovery from the vadose zone. Several experiments were canceled after acoustic application caused reduced (instead of increased) permeability.

In summary, enhanced oil recovery tests and a small LNAPL demonstration in the Netherlands have been completed. DNAPL applications are currently conceptual in nature. The theoretical reasons for acoustic augmentation of electrokinetics and electroacoustics have been proposed. Direct acoustic enhanced removal mechanisms have also been proposed, but are currently unproven especially for DNAPL.

Future efforts should focus on the operative mechanism and on overcoming limitations. Higher efficiencies are needed for successful NAPL remediation. Finally, the operative mechanisms need to be sufficiently understood to allow proper design and implementation on a field scale. Examination of the acoustic enhancement of electrokinetic and electro-osmotic processes on a field scale may also be useful.

Summary

Several technologies are available for direct application of energy into subsurface systems. In general, the added energy helps overcome NAPL related mass transfer and physical/chemical limitations, enhancing the baseline extraction methods. Based on the work to date, resistive heating, radio frequency heating, and electro-osmosis are the most mature direct energy methods and have potential applicability. The technology selection should consider site specific issues such as target contaminant properties, site lithology, and moisture content. For example, for sites with SVOCs, the higher costs of RF heating (versus resistive heating) may be justified to obtain target temperatures in excess of 100 °C.

Acknowledgments

The original unpublished manuscript for this chapter was completed for the Department of Energy, Plumes Focus Area (currently entitled the SUBCON Focus Area) as part of a technical evaluation of enhanced recovery of DNAPL . The authors wish to thank the technical team leaders, Jeff Douthitt and Tom Early, and the technical team members who provided comments on the original draft, Ron Falta, Bernie Kueper, Dave Sabatini, Bob Siegrist, and Kent Udell.

Literature Cited

(1) Dev, H.; Condorelli, P.; Bridges, J.; Rogers, C.; Downey, D. in *Solving Hazardous Waste Problems: Learning from Dioxins;* Exner, J.H., Ed.; ACS Symposium Series No. 338; American Chemical Society: Washington, DC, 1987

(2) Shapiro, A.P. and Probstein, R.F. *Environ. Sci. Technol,* **1993**, 27 (2), pp. 283-291

(3) Beresnev, I.A.; Johnson, P.A. *Geophysics* **1994**, 59 (6), pp. 1000-1017

(4) Webb, S.W. *TOUGH2 Simulations of the TEVES Project Including the Behavior of a Single-Component NAPL;* SAND94-1639/UC-2010 Sandia National Laboratories, Albuquerque, NM, prepared for the U.S. Dept. of Energy, under Contract DE-AC04-94AL85000

(5) Dev, H.; Sresty, G.C.; Bridges, J.E.; Downey, D. in *Proceedings of Superfund '88, HMCRI's 9th National Conference and Exhibition*; Washington, D.C., November 1988

(6) *Innovative Technology Evaluation Report: Radio Frequency Heating, IIT Research Institute;* EPA/540/R-94/527, RREL, Office of Research and Development, US-EPA, Cincinnati, OH, 1995

(7) Phelan, J.M.; Dev, H. presented at the *American Chemical Society I&EC Special Symposium;* Atlanta, GA, September 1995

(8) Edelstein, W.A.; et al. *Environmental Progress,* **1994**, 13 (4), pp.247-252

(9) Jarosch, T.R.; R.J. Beleski; D.L. Faust, *Final Report: In-Situ Radio Frequency Heating Demonstration;* WSRC-TR-93-673, Westinghouse Savannah River Co., Aiken, SC, 1993, prepared for the DOE Office of Technology Development, U.S. Dept. of Energy, under Contract DE-AC09-89SR 18035

(10) *Innovative Technology Evaluation Report: Radio Frequency Heating, KAI Technologies, Inc.;* EPA/540/R-94/528, RREL, Office of Research and Development, US-EPA, Cincinnati, OH, 1995

(11) Buettner, H.M.; Daily, W.D. *J. Environ. Engr.,* **1995**, August, pp. 580-589

(12) Gauglitz, P.A.; et al. *Six-Phase Soil Heating for Enhanced Removal of Contaminants: Volatile Organic Compounds in Non-Arid Soils Integrated Demonstration,* PNL-10184 Battelle Pacific Northwest Laboratory, Richland, WA, 1994, prepared for the Office of Technology Development, U.S. Dept. of Energy, under Contract DE-AC06-76RLO 1830

(13) Ho, S.V.; et al. *Environ. Sci. Technol.,* **1995**, 29 (10), pp. 2528-2534

(14) *Chemical Engineering,* **1995**. Chementator Section, p 23.

(15) Kim, B. C.; S. P. Chauhan; H. S. Muralidhara; F. B. Stulen; B. F. Jirjis *Electroacoustic Soil Decontamination*; 1992 U.S. Patent No. 5,098,538.

Chapter 4

Enhanced Remediation Demonstrations at Hill Air Force Base: Introduction

Philip B. Bedient[1], Anthony W. Holder[1], Carl G. Enfield[2], and A. Lynn Wood[3]

[1]Department of Environmental Science and Engineering, Rice University, 6100 Main Street, Houston, TX 77005
[2]National Risk Management Research Laboratory, U.S. Environmental Protection Agency, 26 West Martin Luther King Drive, Cincinnati, OH 45268
[3]National Risk Management Research Laboratory/Subsurface Protection and Remediation Division at Ada, Robert S. Kerr Environmental Research Center, U.S. Environmental Protection Agency, 919 Kerr Research Drive, Ada, OK 74820

Nine enhanced aquifer remediation technologies were demonstrated side-by-side at a Hill Air Force Base Chemical Disposal Pit/Fire Training Area site. The demonstrations were performed inside 3 × 5 m cells isolated from the surrounding shallow aquifer by steel sheet piling. The site was contaminated with a light non-aqueous phase mixture of chlorinated solvents and fuel hydrocarbons. The technologies demonstrated manipulate the solubility, mobility, and volatility of the contaminants in order to enhance the aquifer remediation over a standard 'Pump-and-Treat' system. Thousands of samples were collected as part of tracer tests, soil flushing demonstrations, and routine characterization efforts. Included in this chapter are the project goals, site information, and details on test cell construction and contaminant distribution. Specific details about each of the technologies can be found in other chapters. An integrated document, the *Guidance Manual for the Extraction of Contaminants from Unconsolidated Subsurface Environments* which will include comparisons of the treatability studies, will be available in the near future.

The *Guidance Manual for the Extraction of Contaminants from Unconsolidated Subsurface Environments* is a research project directed by Rice University's Department of Environmental Science and Engineering. The Guidance Manual project consists of five aquifer remediation treatability studies. In addition, four separately funded Studies that were conducted at Hill Air Force Base (Hill AFB), Utah in 1994 and 1996. The work at this site included 9 demonstrations with many collaborators, funding sources, and funding vehicles. For example, funding sources included SERDP, EPA, AATDF, and Hill AFB. The purpose of the studies was to evaluate innovative technologies for the removal of non-aqueous phase liquids (NAPL) from the saturated and, in some cases, the unsaturated zone. Hill AFB's OU 1 site was chosen because all nine technologies could be demonstrated side-by-side at the location, with similar hydrogeology at a single contaminated site. The studies were conducted in accordance with federal and state regulatory framework so that the results could be incorporated into a treatability manual for the remediation of LNAPL contamination at Hill AFB OU 1 (1, 2). The technologies evaluated include:

Cell 1   air sparging/soil vapor extraction (Michigan Technological University) [SERDP/EPA]

Cell 2   in-well aeration (University of Arizona) [SERDP/EPA]

Cell 3   cosolvent mobilization (Clemson University) [SERDP/EPA]

Cell 4   complexing sugar flush (University of Arizona) [SERDP/EPA]

Cell 5   surfactant solubilization (University of Oklahoma) [SERDP/EPA]

Cell 6   middle-phase microemulsion (University of Oklahoma) [SERDP/EPA]

Cell 7   steam injection (Tyndall Air Force Base, in conjunction with Praxis Environmental and Applied Research Associates) [SERDP/DAF]

Cell 8   single-phase microemulsion (University of Florida) [AATDF/DAF/EPA]

Cell 9   cosolvent solubilization (University of Florida, 1995) [SERDP/EPA]

The major source of funding is indicated in square brackets. Hill AFB provided significant support for all the tests. For some of the technologies, this field application is the first scale-up since the technology was developed in the laboratory.

To facilitate comparisons between the nine technologies being demonstrated, the cell instrumentation, characterization, and basic study methodology were standardized prior to the beginning of field work. Each of the studies was conducted in a test cell designed to separate the test area from the surrounding environment. This was done to facilitate mass balance and performance assessment, as well as to minimize contaminant migration resulting from mobilization and/or solubilization of compounds that are currently not mobile or sparingly soluble (3, 4). Each test cell is approximately 3 m × 5 m, and is constructed with sealable-joint steel sheet pile walls which were driven through the aquifer material and several feet into an underlying clay unit. Further construction/ instrumentation information is discussed below. The demonstrated technologies facilitate contaminant removal through dissolution, emulsification, or mobilization in groundwater or through volatilization and enhanced bioremediation when air delivery systems were used. In some cases, this was the first time that remediation approach had been demonstrated in the field.

Solubilization. Organic NAPL constituents tend to be hydrophobic (water hating) because of their non-polar, non-ionic nature. Consequently, the aqueous solubility of most LNAPL constituents is low. Macromolecules and surfactants behave as small organic particles that can move through the soil because of their hydrophilic exteriors. The interior of these particles is hydrophobic, and they are able to carry significant amounts of NAPL constituents. On a macroscopic scale, there appears to be an increase in solubility due to the presence of surfactants or macromolecules. Cosolvents change the dielectric properties of the fluid, making it less polar and thereby increasing the fluid's ability to dissolve the NAPL constituents. The solubilization technologies include Cell 4 - complexing sugar flush and Cell 5 - surfactant solubilization, Cell 9 - cosolvent solubilization.

Volatilization. Air sparging/soil vapor extraction, in-well aeration, and steam injection all attempt to remove NAPL constituents through volatilization and concurrently augment bioremediation by supplying oxygen to the system. The air sparging/soil vapor extraction technology and the in-well aeration technology are similar in principle, but differ in mode of application. Both technologies attempt to volatilize and remove volatile constituents that are dissolved in the groundwater using injected air. Air sparging uses multiple wells accompanied by vapor extraction from the vadose zone; in-well aeration volatilizes the constituents within a single borehole and then vents the gas to the atmosphere via the well casing. Steam injection increases the volatility, and thus the rate of removal, of NAPL constituents over simple air sparging by increasing the subsurface temperature. The volatility technologies include Cell 1 - air sparging/soil vapor extraction; Cell 2 - in-well aeration; and Cell 7 - steam injection.

Mobilization. Because of high interfacial tension between the groundwater and NAPL residual phase, LNAPL is trapped in the spaces between soil grains. Surfactant and cosolvent mobilization and single phase microemulsification technologies attempt to make the NAPL move with the solution pumped through the contaminated zone. Adding surfactants and/or cosolvents to the groundwater can significantly reduce the interfacial tension between the NAPL and the groundwater. The reduced interfacial tension allows the NAPL and the groundwater to move as a single, emulsified phase, dramatically increasing NAPL mass removal efficiencies. In some cases, the organic phase moves as a separate bulk phase apart from the aqueous phase. The mobilization technologies include Cell 3 - cosolvent mobilization; Cell 6 - middle-phase microemulsion; and Cell 8 - single phase microemulsion. See the following chapters for more details on the technologies.

Site Background

Hill AFB is in northern Utah, approximately 40 km north of Salt Lake City and five miles south of Ogden. The portion of the base selected for the treatability studies is known as Operable Unit 1 (OU 1), and is located near the northeastern boundary of Hill AFB. The disposal sites at OU1 have been designated as: Chemical Disposal Pits (CDPs) 1 and 2; Landfills (LFs) 3 and 4; Fire Training Areas (FTAs) 1 and 2; Waste Phenol/Oil Pit (WPOP); and Waste Oil Storage Tank (WOST). The treatability study test cells are adjacent to CDPs 1 and 2 (Figure 1 and Figure 2). The CDPs were industrial liquid waste disposal sites in operation from 1952 through 1973. Wastes (principally petroleum hydrocarbons and spent solvents) were burned periodically at these sites (2).

Prior to the beginning of the enhanced remediation demonstrations at Hill AFB, Montgomery Watson, Inc. had performed site characterization work for the remedial investigation. Their work included installing monitoring wells, collecting soil cores, performing slug tests, and characterizing the geology of the area. Much of the background information presented in this section was adapted from their reports (1, 2).

Geologic and Hydrogeologic Setting. About 6 km to the east of Hill AFB, the Wasatch Mountains rise abruptly to an elevation of over 2900 meters above mean sea-level (amsl). The Great Salt Lake is approximately 20 km west of Hill AFB at an elevation of 1280 meters amsl. Hill AFB is on a terrace of the Weber Delta approximately 90 m above the Weber River valley floor. The valley runs east-west immediately north of Hill AFB. The elevation of OU 1 is approximately 1500 meters amsl.

The sedimentary formation which contains the shallow aquifer on base consist of two sedimentary units. A possible third, deeper unit has been penetrated by only one boring; therefore, its lateral extent is unknown. The subsurface stratigraphy at OU 1 can be characterized as follows:

Upper Sand and Gravel Unit: An upper unit consisting of fine to coarse, clean to silty sand interbedded with gravel and cobbles that ranges in thickness from 0 to 19 m and averages 9 m.

Silty Clay Unit: An intermediate unit primarily consisting of silty clay interbedded with fat clay and silt, and containing thin stringers (tenths of centime-

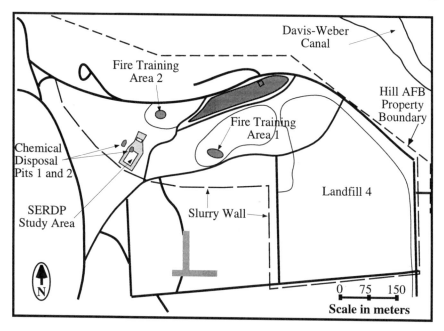

Figure 1: Hill Air Force Base Operable Unit 1 (OU 1).
Source: ADAPTED FROM REF. 1.

ters to a few centimeters thick) of very fine sand. The unit is approximately 60 m thick and appears to be saturated, with intermittent saturated sand stringers from its top to as deep as it has been penetrated by drilling. A northwest trending gravel and sand-filled paleo channel within the top of the silty clay unit has a saturated thickness of up to two meters.

Deeper Sand Unit: A possible deeper unit of unknown thickness consisting of clean sand with occasional stringers and interbeds of silty to fat clay.

The presence of the silty clay unit played a major part in site selection, allowing for the construction of the cells to hydraulically isolate the treatability studies. The saturated portion of the upper sand and gravel phreatic aquifer has a horizontal hydraulic conductivity ranging from 10^{-1} to 10^{-2} cm/sec based on aquifer test data, and from 10^{-2} to 10^{-5} cm/sec based on slug test data. The horizontal hydraulic conductivity of the underlying clay unit ranges from 10^{-4} to 10^{-5} cm/sec based on slug test data. The calculated average linear horizontal velocity of on-base groundwater ranges from 3.0×10^{-4} to 4.5×10^{-3} cm/s in the upper sand and gravel unit, and 4.0×10^{-7} to 4.8×10^{-5} cm/s in the clay unit. The groundwater elevation in the study area fluctuates between 5.5 to 7.5 m bgs. Inside several of the cells, the water levels were raised for the duration of the treatability studies.

Disposal History. The disposal history of OU 1 indicates that bulk quantities of liquid waste were disposed of and periodically burned at the Chemical Disposal Pits (CDP's). A distinct black and oily layer was observed in all boreholes drilled for

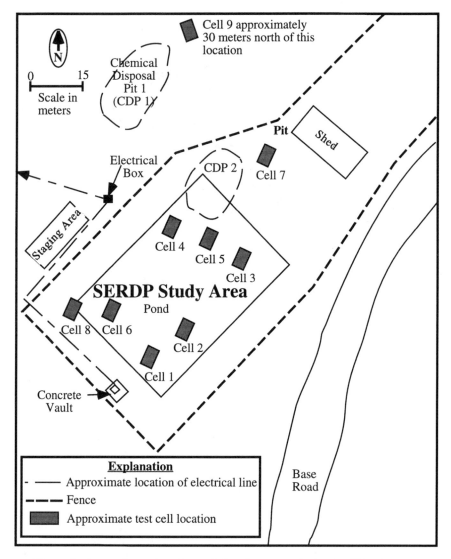

Figure 2: Approximate locations of the nine test cells in the SERDP treatability study area.
Source: ADAPTED FROM REF. 1.

these studies, indicative of free and residual phase LNAPL. The thickness of this layer ranged from 3 to 6 meters, ending at the interface between the sand/gravel unit and the silty clay unit. Staining was observed in the silty clay unit, generally extending only a few inches.

Contaminant composition varied from borehole to borehole. Jet fuel is the primary component of the LNAPL present at OU 1. Other contaminants found in the LNAPL include volatile organic compounds (VOC's), base/neutral and acid extractable compounds (BNAE's), polychlorinated biphenols (PCB's), pesticides, dioxins, and furans. Soil sample analysis yielded above background concentrations of chromium, copper, lead, mercury and zinc. Arsenic and PCB's have also been detected in the surface soils.

Contaminant Distribution. Based on the data available prior to the initiation of the studies, the principal contaminant sources at OU 1 are the contaminated soils in the vadose zone underlying the disposal sites, the residual- and free-phase LNAPL in the capillary fringe, and the contaminated aquifer matrix soils at and downgradient of CDPs 1 and 2, FTA 1, and the western part of Landfill 3. The contaminants include VOCs, BNAEs, PCBs, dioxins, and furans. The following sections discuss contaminant types and distribution in the various media at OU 1, and focus on the area adjacent to CDPs 1 and 2 where the treatability studies were performed.

LNAPL. Based on the history of the OU 1 disposal sites, bulk quantities of liquid wastes were disposed and periodically burned at the CDPs. During previous drilling operations, a distinct "black or oily" layer with a strong hydrocarbon odor was observed in most boreholes associated with the CDPs. The layer was first encountered just above and/or below the water table, and ranged from approximately 0.3 to 1.5 meters in thickness, and generally ended at the contact between the sand and gravel unit and the silty clay unit. Stained soil was observed in the silty clay unit in several boreholes, but the staining extended less than two feet into this unit.

In several soil borings associated with the CDPs a distinct fuel layer was observed above the black layer. The layer consisted of soil visibly coated with a substance that had a strong fuel odor and caused a sheen on the soil particles, but did not discolor or stain the soil. The fuel was observed in the sand and gravel unit 3.5 to 8.2 meters bgs. The apparent thickness of the fuel layer varied from approximately 1 to 3 meters. This fuel layer probably represents residual LNAPL saturation in the soil above the water table. This layer may have formed as ground-water levels fluctuated and contaminants were smeared in the capillary fringe. Residual LNAPL was also observed below the water table.

Based on laboratory analyses, the LNAPL is composed primarily of jet fuel; however, other contaminants that have been disposed at OU 1 have solubilized and accumulated in the LNAPL due to co-solvency. VOCs, BNAEs, pesticides, PCBs, dioxins, and furans have been detected in the LNAPL associated with CDPs 1 and 2. The apparent free phase LNAPL thickness based on monitoring well observations ranges from 0 to 0.3 meters in the area of the proposed treatability studies. This thickness does not include residual LNAPL.

Soil. Several soil borings have been taken in the CDP 1 and CDP 2 area. The areal extent of contaminated soil appears to increase between 4 and 8 m bgs, and then to decrease below the water table. In addition, the concentrations of individual contaminants in specific borings are greater in the 4 to 7 m bgs interval than in the 7 to 8 m bgs interval. High concentrations of fuel hydrocarbons and chlorinated hydrocarbons in the subsurface soils at the CDPs is consistent with the types of wastes disposed in this area.

Groundwater. During the Phase I and II Remedial Investigation and the quarterly monitoring program, 27 wells in the CDP area on the Base and on the hillside downgradient of the CDPs were sampled. Generally, the types of compounds identified in the groundwater at the CDPs are consistent with the history of disposal activities in the area. The majority of the wastes disposed in this area were spent chemical solvents and petroleum hydrocarbons.

Chlorinated hydrocarbons were the most common groundwater contaminants detected in the CDPs during the remedial investigation process. The VOC detected at the highest concentration and with the widest distribution is 1,2-DCE. Chlorinated benzene compounds also were detected over a wide area and in relatively high concentrations, although their distribution is much more limited than that of 1,2-DCE. The chlorinated benzene compounds detected include chlorobenzene, 1,4-dichlorobenzene, 1,2-dichlorobenzene, and 1,2,4-trichlorobenzene. PCBs, dioxins/furans, and chlorinated pesticides were detected in several ground-water samples. However, these compounds adsorb readily to soil particles and solubilize in LNAPL; consequently, suspended soil particles or a small amount of emulsified LNAPL in the ground-water sample could account for the observed concentrations of these compounds. Table I summarizes the LNAPL, soil, and groundwater analyses performed as part of the site characterization for the remedial investigation.

Test Cell Layout / Instrumentation

Each of the nine treatability study test cells has a similar layout of injection and extraction wells, multilevel groundwater samplers, piezometers, and monitoring wells. A schematic of the standard test cell is shown in Figure 3. This standard layout was used in all test cells to allow the pre- and post-treatment tracer tests to be conducted using a common configuration. The baseline layout without modification was sufficient for performing the six treatability studies based on flushing, namely: cosolvent mobilization, complexing sugar flush, surfactant solubilization, surfactant middle phase microemulsion, single phase microemulsion, and cosolvent solubilization. Modified layouts and additional instrumentation for the remaining three treatability studies (steam injection, in-well aeration, and air sparging/soil vapor extraction) are shown or described in the relevant chapters.

Test Cell Construction

The first test cell was constructed in the fall on 1993 and the demonstration performed in 1994. The remaining eight test cells were installed during fall 1995 through spring 1996, and the characterization and technology demonstrations took place during the summer and fall of 1996. Detailed cell construction procedures are described in Phase I and Phase II Work Plans (1, 2). Each of the studies were conducted in an isolation cell to separate the test area from the surrounding OU 1 environment, to mitigate contaminant migration resulting from mobilization or solubilization of contaminants that currently are not mobile or are sparingly soluble in ground-water under present site conditions, and to aid in mass balance calculations for the technology evaluation/comparison. Figure 2 shows the area near CDPs 1 and 2 selected for the treatability studies and the location for each of the nine test cells. This site was the location of a temporary holding pond that was used until the spring of 1994 to combine effluent from the OU 1 dewatering system. The holding pond was an earthen bermed area lined with HDPE and open to the air. In order to use this site, rainwater in the pond, accumulated sediments, and the high density polyethylene liner were removed and the pond was backfilled with clean fill. The site was brought to grade prior to installation of the test cells.

Figure 3 shows the base configuration for each test cell. Extraction wells, injection wells, multi-level ground-water samplers, piezometers, and monitoring wells were installed adjacent to and within each test cell to inject treatment solutions, extract effluent, monitor the efficiency of the system, and/or monitor any ground-water leakage from the test cells. The multi-level samplers, wells, and piezometers were installed with either a cone penetrometer truck (CPT) or with a standard hollow stem auger. The CPT was the first choice, because it has a smaller impact on the subsurface geology; however in many cases, the CPT was inappropriate because of large cobbles found in the Upper Sand and Gravel Unit.

Table I. Summary of Site Characterization Results for LNAPL, Soil, and Groundwater at Hill AFB.

Contaminant	LNAPL (μg/kg) (U1-004)	Soil (μg/kg) (Maximum)	Groundwater (μg/l) (Maximum)
Trichloroethylene	not tested	40,000	2,300
Tetrachloroethylene	38,000	9,100	150
1-2-Dichloroethylene	87,000	4,200	42,000
1,1-Dichloroethylene	not tested	790	not tested
1,1,1-Trichloroethane	92,000	8,100	1,500
Chlorobenzene	<13,000	2,000	4,900
1,2-Dichlorobenzene	2,700,000	170,000	190
1,4-Dichlorobenzene	not tested	21,000	130
Total Polychlorinated Biphenyls	190,000	8,300	150

SOURCE: Adapted from ref. 2.

Test Cell Injection/Extraction Wells. A bank of four injection wells and a bank of three extraction wells were installed into each cell in the locations as shown in Figure 3. The injection wells are spaced evenly along a line parallel to the short side of the cell, approximately 0.3 m from the end of the cell. At the opposite end of the cell is an evenly spaced bank of three extraction wells, as shown in Figure 3. The injection/extraction system for the test cell was designed to optimize uniform flow. The same layout was used in all studies to maintain comparability.

The injection/extraction wells were installed with stainless steel screens and PVC risers. During the installation, soil cores were collected at selected well locations for NAPL evaluation. The injection and extraction wells were designed and constructed in the same manner.

Test Cell Multilevel Samplers. Each cell has a minimum of a three by four matrix of multilevel samplers (MLS) evenly spaced between the banks of injection and extraction wells as shown in Figure 3. Each multilevel sampler is designed to collect groundwater or soil gas samples from five or more levels within the formation. Each MLS consists of 5 or more sampling ports, spaced 0.3 to 0.9 m (1 to 2.5 feet) apart, vertically distributed through and adjacent to the target treatment zone. The multilevel samplers were used during the tracer tests and treatability studies to collect groundwater and soil gas samples for analysis. The results were used to monitor flow characteristics and composition of the fluids of the tracer tests and the technology demonstration test and to adjust the treatment process, if necessary. Soil samples were collected at selected MLS locations to characterize the pretreatment soil conditions in the test cell.

Test Cell Monitoring Wells. A pair of 5 cm (2 inch) diameter, nested monitoring wells were installed down gradient, which was the north side of each cell. Each nested pair of monitoring wells consists of one well screened across the sand and gravel unit, the other screened in the clay unit. Each cell also has a minimum of two 1.25 cm (0.5 inch) diameter piezometers. The piezometers are located along one side of the test cell and evenly spaced between the banks of injection and extraction wells.

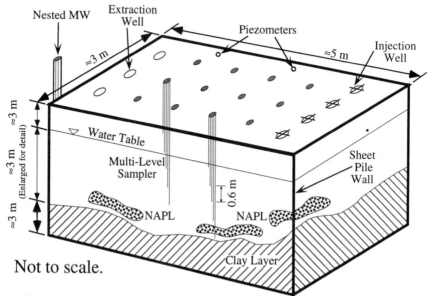

Figure 3: Schematic representation of a treatability study test cell.

Tank Farm. Several storage tanks were necessary for the operation of the treatability studies. To ensure a steady supply of water for the flushing demonstrations and tracer tests, water was stored prior to mixing with tracers or flushing agents. Separate tanks were necessary to ensure adequate mixing of the tracers or flushing agents. Some of the wastes were transported off site for disposal or incineration. The remainder of the extracted water was stored and tested before disposal at the Hill AFB Industrial Wastewater Treatment Plant (IWTP), to ensure that the IWTP did not receive a slug of highly concentrated waste that could not be treated properly. Thus, during the height of flushing demonstrations, there was an influent tank farm (consisting of three 75,000 L tanks and three 23,000 L tanks), and an effluent tank farm (consisting of eight 75,000 L tanks and two 23,000 L tanks), along with two 23,000 L water supply tanks and related plumbing.

Test Cell Characterization

Characterizing the spatial distribution of contaminants within each of the test cells before and after remediation is a critical activity for evaluating the performance of the technologies. The performance of the innovative technologies will be evaluated based on the change in the amount of NAPL mass within each of the test cells, and on the amount of mass remaining in the cell after treatment. If properly designed, partitioning tracer studies can give an estimate of the distribution of NAPL within a test cell, as well as the total mass in the cell (5). NAPL distribution measures were an objective of the tracer tests performed as part of this project. Tracer studies were performed to permit estimating the total mass of NAPL pre- and post-remediation as one measure of the performance of the remediation technology.

Pre- And Post-Treatment Sampling And Analysis. Pre- and post-treatment soil and post-treatment groundwater samples were collected and analyzed to characterize the contaminant composition, distribution, and mass in the cells after treatment, and to evaluate the effectiveness of each treatment technology for removing individual NAPL constituents. The evaluation will based on total contaminant mass removed versus total pore volumes flushed during the test and a knowledge of the mass of contaminant of time zero. For consistency, all soil and groundwater samples were analyzed for the following twelve constituents: trichloroethene, o-xylene, m-xylene, undecane, 1,2-dichlorobenzene, naphthalene, 1,1,1-trichloroethane, benzene, toluene, ethylbenzene, decane, and 1,3,5-trimethylbenzene.

Soil Sampling. During the pre-treatment soil sampling, eight soil borings per cell were collected at selected well or MLS locations. Soil samples were collected at intervals of 15 to 60 cm to estimate the distribution of several target constituents, the NAPL composition, spatial distribution, and mass in the soil for each cell. Average spatial mass estimates for the indicator parameters were calculated using the results. During post-treatment soil sampling only four to six soil borings were possible in each cell due to space constraints. Soil samples provide point information on contaminant distribution and composition, but do not give good estimates of total mass in the swept volume being remediated. By defining the statistical distributions of the data with parameters such as means and standard deviations, statistical tools such as the variance of means and trend analyses permit the determination of the significance of the difference achieved by treatment. Most of the samples were analyzed at Michigan Technological University to permit a consistent laboratory technique and improve study to study comparisons.

Groundwater Sampling. Groundwater samples collected before and after treatment were analyzed for target NAPL constituents in order to evaluate the changes in groundwater quality resulting from a treatment technology, and to determine the effect of the treatment action on individual contaminant partitioning between the soil and groundwater at equilibrium. Statistical analysis of the concentrations of the indicator compounds provides an understanding of the spatial distribution and total mass of each selected indicator compound after treatment.

Tracer Tests

Partitioning tracer tests (PTTs) were performed in all nine test cells both before (pre-PTT) and after (post-PTT) the technology demonstrations. These tests are designed to determine the spatial distribution and the total volume of NAPL within each cell. The tracer tests include a laboratory portion and a field portion. The laboratory tests were used to evaluate several tracers for use in the field tests, including: ethanol, methanol, hexanol, bromide, 2,2-dimethyl-3-pentanol, n-pentanol, and 6-methyl-2-heptanol.

During the partitioning tracer test, a small volume of solution containing low concentrations of both conservative and partitioning tracers was pumped through the cell. The conservative tracer will pass through the system unaffected by the NAPL content of the cell. The partitioning tracers will partition into the NAPL phase when they encounter it, and partition back out of the NAPL after the main mass of the tracer pulse has passed, resulting in a chromatographic separation between the partitioning and conservative tracers. This chromatographic separation is reflected in concentration breakthrough curves (BTCs) developed for the various sampling points in each cell (Figure 4). The laboratory evaluation of tracers is necessary to determine partitioning coefficients of the tracers between water and NAPL. If the separation is too small, the accuracy of the method is compromised, however if it is too large, the tracers emerge from the cell more slowly, and a longer test is more costly (5, 6). Several partitioning tracers were used to ensure proper chromatographic separation of breakthrough curves, and to improve estimates of NAPL saturation.

Pre-treatment tracer test results were also used to estimate the effectiveness of "Pump-and-Treat" technologies for removing NAPL and to compare the innovative treatment technologies to conventional "Pump-and-Treat" technologies. Post-treatment water samples are used to document the concentrations of individual constituents in groundwater at equilibrium after treatment.

Laboratory-Based Tracer Tests. Prior to the field work, the tracers were evaluated in the laboratory using static partitioning tests (batch tests) and dynamic partitioning tests. The results of these laboratory tests were used to select partitioning tracers for use in the field. Static partitioning tests were used to select reactive tracers for further testing based on the estimated retention time. The equilibrium distribution of tracer between the aqueous and NAPL phases was measured. The static partitioning coefficient is the ratio of tracer concentration in NAPL to tracer concentration in the aqueous phase.

Dynamic partitioning tests were used to estimate NAPL content under dynamic conditions. Breakthrough curves were constructed by plotting concentration in the effluent versus time for the tracers. Mean residence times are calculated by integrating the breakthrough curves with respect to time. Partitioning coefficients were determined by static liquid-liquid tests, and confirmed with column tests using Hill AFB sediments.

Tracers for use in the field tests were chosen based on their partitioning behavior between water and NAPL from the test site. Other desirable tracer properties include: 1) non-toxic, 2) non-hazardous, 3) non-degrading, 4) low volatility, 5) reasonable cost and availability, and 6) easily quantifiable in the presence of NAPL constituents (Annable, M.D., P.S.C. Rao, W.D. Graham, K. Hatfield, and A.L. Wood, Use of partitioning tracers for measuring NAPL: Results from a field-scale test, *J. Environ. Eng.*, in press, 1997).

Field Tracer Tests. All tracer tests were designed and conducted according to procedures established by Jin and Annable (5, 6). A minimum of three tracers (one non-partitioning and two partitioning) were used for the tracer tests. Tracer tests also act as a pump-and-treat system, the data from which was used to establish benchmark conditions. The tracer tests were performed using the following basic procedure: 1) establish steady-state flow conditions in the cell for 2 days to ensure adequate system stability; 2) inject 0.1 to 0.2 pore volumes of tracer; 3) sample at MLS points and extraction wells. The sampling schedule typically was hourly for the first 12 - 24 hours, slowing to every 4 - 8 hours for up to 10 days, until the tracers have been adequately recovered. Hydraulic gradients and flow rates were monitored continuously and recorded during each test.

Breakthrough curves for the reactive tracers were used to estimate the volume and spatial distribution of NAPL in each test cell. A minimum of three methods are used to calculate the NAPL distribution. First, the integrated mass of NAPL in the swept volume is calculated by the method of moments. Second, an inversion technique is used with a solution to the 1-D advection–dispersion equation utilizing a method of superposition to account for spatial variability along the flow path. The third method will be an inversion technique that incorporates the code developed by Dr. Gary Pope at the University of Texas. Calculations can be made for each of the MLS points and the extraction well.

A fourth analysis method has been developed, also based on the 1-D advection–dispersion equation. This method involves fitting the data at each of the MLS points to the 1-D solution, using 4 fitting parameters. Data for each tracer at each MLS point are required to fit the curve using the same values for all parameters except the velocity. The ratios of velocity are then used to determine the retardation factors and thus the NAPL saturations. By estimating the values of the 4 parameters

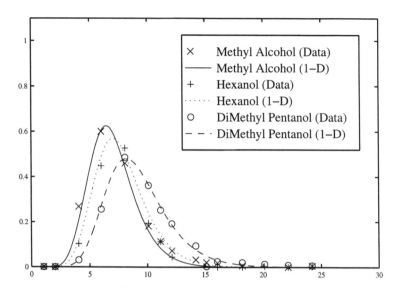

Figure 4: Typical Partitioning Tracer Test Breakthrough Curve.

over the test cell, and applying the 1-D equation at each point in the cell, an estimate of the concentrations at each point in time and space for the entire cell may be calculated. These estimates can then be used for data visualization, including 3-D animations of the tracer test. Figure 4 shows a typical set of breakthrough curves at an MLS point, and the best fitting 1-D advection–dispersion solution for those three curves. This method will be used for some of the tests.

The method of moments will also be used in combination with head measurements to determine the water-filled porosity and saturated hydraulic conductivity. These calculations will be made for each of the MLS sampling locations and the extraction wells.

Considerations for Post-Treatment Tracer Tests. The composition and volume of residual NAPL were altered during the technology demonstration tests. Because of this, a tracer with a higher partitioning coefficient was generally added for the post-treatment tracer tests. Three tracers (methanol, hexanol, and 2,2-dimethyl,3-pentanol) were used in all test cells. This provides a consistent method for comparing and contrasting results from each of the test cells.

Performance Criteria

The innovative technologies are currently being compared against each other and conventional "Pump-and-Treat" is serving as the reference technology. The pre-PTT performed in each test doubles as a short "Pump-and-Treat" demonstration. This short test serves as a measure of "Pump-and-Treat" performance. Three performance measures have been identified for evaluating and comparing the performance of the nine remedial technologies. These measures are:

- Pre- and post-treatment soil sampling, analysis and statistical evaluation using indicator compounds to evaluate the removal of various constituents and families of contaminants.

- Estimating total contaminant removal by monitoring and integrating the total amount of contaminant removed in each process stream (e.g., extraction fluid, soil gas, fugitive emissions).
- Estimating pre- and post-treatment NAPL mass, and therefore mass removal, via partitioning tracer tests.

Summary

Hill AFB provided a unique opportunity for side by side testing of innovative technologies at an actual field waste site. The testing, which consisted of extensive pre-test characterization, soil flushing technology demonstration, and post-test characterization, was performed inside sheet pile cells. The cells were designed to isolate each demonstration, and to minimize any spreading of contamination. Treatability studies were performed on nine technologies, which used solubilization, mobilization, and volatilization techniques to remove LNAPL from the subsurface.

During the course of the cell characterization and technology demonstrations, over 20,000 soil or groundwater samples were collected for analysis at each cell, for a total of more than 100,000 samples for the entire project. More than 60 individual investigators and regulatory personnel were involved with the tests. The analysis of data and the comparison of technologies is ongoing. Results from each treatability study are given in following chapters.

Acknowledgment/Disclaimer

This study was funded by the Strategic Environmental Research & Development Program (SERDP), which is a collaborative effort involving the U.S. EPA, U.S. DOE, and U.S. DoD. Supplemental financial and technical assistance was provided by the U.S. EPA's National Risk Management Research Laboratory (NRMRL) and the Environmental Restoration and Management Division, Hill Air Force Base, Layton, Utah. Although SERDP, EPA and Air Force funds were used to support this study, this document has not been subjected to peer review within the agencies, and the conclusions stated here do not necessarily reflect the official views of the agencies, nor does this document constitute an official endorsement by the agencies.

Literature Cited

1. Montgomery-Watson, 1995. *Phase I Work Plan for Eight Treatability Studies at Operable Unit 1*. Prepared for Hill AFB.
2. Montgomery-Watson and Dynamac, 1996. *Phase II Work Plan for Eight Treatability Studies at Operable Unit 1*. Prepared for Hill AFB.
3. Starr, R. C.; Cherry, J. A.; Vales, E. S. *45th Canadian Geotechnical Conference,* **1992**.
4. Starr, R. C.; Cherry, J. A.; Vales, E. S. *Technology Transfer Conference,* **1993**.
5. Jin, M.; Delshad, M.; McKinney, D. C.; Pope, G. A.; Sepehrnoori, K.; Tilburg, C.; Jackson, R. E. *American Institute of Hydrology Conference,* **1994**.
6. Annable, M. D.; Rao, P. S. C.; Hatfield, K.; Graham, W. D.; Wood, A. L. *Second Tracer Workshop,* **1994**.

Chapter 5

Field Demonstration Studies of Surfactant-Enhanced Solubilization and Mobilization at Hill Air Force Base, Utah

Robert C. Knox[1], Bor Jier Shau[2], David A. Sabatini[1], and Jeffrey H. Harwell[3]

The Schools of [1]Civil Engineering and Environmental Science and [3]Chemical Engineering and Materials Science, The Institute for Applied Surfactant Research, University of Oklahoma, Norman, OK 73019
[2]Mantech Environmental Research Services Corporation, 919 Kerr Research Drive, Ada, OK 74820

Surfactant-enhanced subsurface remediation can dramatically improve contaminant removal rates compared to the traditional pump-and-treat technology. Surfactants can be used to significantly enhance the solubility (solubilization) of non-aqueous phase liquid (NAPL) constituents, or they can be used to reduce interfacial tensions thereby mobilizing the NAPL (mobilization). Both the solubilization and mobilization mechanisms were used to remediate separate portions of a NAPL-contaminated aquifer at Hill AFB, Utah. The demonstrations were conducted in cells contained by steel sheetpiling driven into an underlying impermeable layer. The solubilization demonstration cell showed excessive leakage through the sheetpiling; hence, the surfactant (Dowfax 8390) was not flushed through the entire cell. In spite of less than complete flushing of the cell, contaminant extraction in the solubilization cell was as high as 58% with ten pore volumes of surfactant flushing. In the mobilization cell, two surfactant solutions (Aerosol OT and Tween 80) were injected along with calcium chloride. The surfactant-mobilized NAPL had a higher viscosity than aqueous fluids resulting in reduced hydraulic conductivities and mounding of the solution. In under seven pore volumes, the average contaminant removal for the mobilization cell exceeded 90%. Flushing with water alone would have extracted less than 1% of the contaminant mass in the same time frame. The contaminant removal rates derived from pre- and post-demonstration soil cores are in concert with the contaminant mass removed in the extraction wells and generally agree with the pre- and post-demonstration partitioning tracer tests. The mobilization system was much more efficient than solubilization with both systems being much more efficient than water alone. By contrast, the solubilization system was much easier to design and implement. These demonstrations thus illustrate the exciting potential for surfactants to dramatically improve pump-and-treat remediation of residual oil.

Background

Water-based pump-and-treat remediation has proven inefficient for remediating residual NAPL contamination. Cost overruns of 80% and cleanup times as much as three times longer than original estimates are typically reported (1). Several factors have been identified that contribute to the inefficiency of pump-and-treat remediation, including: (1) diffusion limitations for contaminants from low conductivity zones to high conductivity zones; (2) hydrodynamic isolation, such as dead end zones; (3) rate-limited desorption of contaminants from solid surfaces; and (4) liquid-liquid dissolution of residual NAPLs (2,3,4,5). It is thus apparent that innovative technologies are necessary for treating NAPLs as pump-and-treat alone will be extremely inefficient.

Chemical amendments that have been suggested for expediting pump-and-treat remediation include the following: complexing agents, cosolvents, surfactants via solubilization and mobilization, oxidation-reduction agents, precipitation-dissolution reagents, and ionization reagents (6). Surfactant-enhanced subsurface remediation has received increasing attention as a viable technology for remediating NAPL source zones (7). Surfactant enhancements can result from solubilization or mobilization. Surfactant micelles have a hydrophilic exterior and a hydrophobic interior. The greater the surfactant concentration above the critical micelle concentration (CMC) and the more hydrophobic the contaminant, the larger the solubility enhancement anticipated; this is referred to as the solubilization mechanism.

Surfactant mobilization (formation of a middle phase microemulsion with concomitant ultra-low interfacial tensions -- IFTs) results in bulk displacement of the residual oil phase. A middle phase system occurs on the continuum between aqueous phase micelles (described above) and reverse micelles (having a hydrophobic exterior and a hydrophilic interior, and located in the oil phase). The transition between these three phases can be realized by various methods, including variations in salinity, temperature or variations of concentrations in mixed surfactant / hydrotrope systems (8). Both of these approaches were demonstrated at Hill AFB, Utah, in the summer of 1996.

Surfactant Demonstration Studies

Hill AFB, located north of Salt Lake City, UT, has been the site of military activities since 1920. The particular area of interest is Operable Unit 1 (OU1), which includes several landfills, chemical disposal pits and fire training areas. Principal contaminants of concern are light NAPLs (LNAPLs, e.g., jet fuel, light lubricating oils), chlorinated VOCs, PCBs, etc. The soil of interest is predominantly sand and gravel, with less than five percent silt, clay and organic content.

A series of test cells were installed at OU1 for the purpose of demonstrating nine innovative remediation technologies, including the surfactant solubilization and mobilization cells used by our group. As shown in Figure 1, each 3m x 5m cell consisted of sheetpiling driven down to a lower confining layer (8 to 10 m). Each cell had three extraction wells and four injection wells to establish the horizontal flow field. A grid of multilevel samplers allowed three dimensional characterization during the studies; sixty sampling points result from the three by four grid with five depths at each grid point.

The three dimensional contaminant distribution in each cell was assessed via pre- and post-demonstration soil corings and partitioning tracer tests (PTTs). Pre- and post-demonstration soil cores were taken at eight locations across each cell. At each location, cores were taken over 2-foot depth intervals and composited. For the PTTs, known masses of conservative tracer (e.g., bromide) and partitioning alcohols (e.g., hexanol, 2,2-dimethyl-3-pentanol) were injected into the cell, followed by water. The method of moments can be used to quantify the chromatographic separation of the conservative and partitioning tracers; the distribution coefficients for the partitioning tracers with the NAPL can then be used to calculate NAPL residual saturation in the cell (9).

One cell (Cell 6) was used to demonstrate surfactant-enhanced solubilization while a separate cell (Cell 5) was used to demonstrate surfactant-enhanced mobilization. The sequence of activities of each demonstration study were as follows: 1) pre-demonstration soil coring, 2) pre-demonstration PTT, 3) pre-demonstration equilibrium water samples, 4) pre-demonstration water flush (2 pore volumes), 5) demonstration surfactant flush (mobilization = 6.6 pore volumes, solubilization = 10 pore volumes), 6) post-demonstration water flush (5 pore volumes), 7) post-demonstration PTT, 8) post-demonstration equilibrium water samples, and 9) post-demonstration soil coring. Water levels and flow rates were measured periodically for the injection and extraction wells during the tracer tests and surfactant flushes. The operating conditions for both tests are listed in Table 1.

Surfactant-Enhanced Solubilization

Laboratory solubilization studies were conducted using site soils and candidate surfactants. The batch studies evaluated solubilization potential, sorption level and susceptibility to precipitation or phase separation for each of the surfactant solutions. Based on theses studies, Dowfax 8390, Steol CS-330 and Witconol SN-120 were

52

Test Cell Configuration

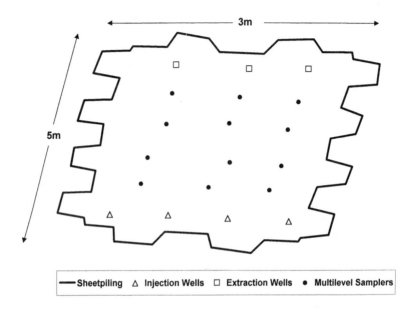

Figure 1. Configuration of test cells.

Table 1: Operating Conditions for Surfactant Demonstration Cells

Location	Activity	Average Saturated Thickness (m)	Treatment Zone Width (m)	Treatment Zone Length (m)	Treatment Zone Pore Volume (L)	Average Head Across Cell (m)	Average Influent Flowrate (lpm)	Average Effluent Flowrate (lpm)
	Pre-PTT	2.62			3404	0.22	2.57	2.69
Cell 6	Demonstration	2.80	2.70	3.69	3113	0.33	2.61	2.46
	Post-PTT	4.33			7617	0.06	5.68	6.02
	Pre-PTT	5.14			9403	0.37	6.70	6.89
Cell 5	Demonstration	5.41	2.70	3.98	8216	2.32	6.06	6.02
	Post-PTT	5.21			8625	0.47	6.25	6.32

selected for further evaluation in column studies (with Witconol SN-120 having the highest solubilization of the NAPL phase). In column studies the Steol CS-330 and Dowfax 8390 outperformed Witconol SN-120, illustrating the importance of column studies in the surfactant screening process. Based on these results, and its successful use in a previous field study (10), Dowfax 8390 was chosen for the field demonstration.

Prior to the demonstration studies, each cell was tested for leakage. The sheetpiling enclosing Cell 6 was shown to have an excessive leakage rate (> 0.3 pore volumes per day). Exterior grouting around the cell did not significantly reduce the leakage rate. Hence, for the demonstration study, surfactant solution was injected into only three of the four injection wells; water was injected into the fourth well located on the leaky side of the cell to provide a buffer zone. This effectively reduced the treatment zone subjected to supra-micellar surfactant concentrations by 25%.

Contaminant removal for the cell was assessed by comparing pre- and post-demonstration soil cores, pre- and post-demonstration PTTs and by calculating contaminant mass in the effluent waste stream. Core recovery for a given interval was sporadic and the depth intervals for the pre- and post-demonstration cores did not coincide. In addition, at least three of the post-demonstration soil cores were taken from zones that may not have been flushed with the full strength surfactant solution due to dilution from water in the fourth injection well.

Figure 2 is a plot of the treatment zone soil contaminant concentrations and removal percentages for just those contaminants showing a positive percent removed. The removal percentages range from 44% (DCB) to 67% (toluene) with an arithmetic average removal of over 58%.

From the pre-solubilization PTT data, hexanol showed poor recovery (< 75%) and anomalous breakthrough characteristics; hence, hexanol results were not used in analyzing the PTT results. The mass recovery rates for the bromide and 2,2-dimethyl-3-pentanol tracers were exceptionally good (> 90%).

For the post-solubilization PTT, calculations using the method of moments with bromide and 2,2-dimethyl-3-pentanol showed only a marginal decrease from pre-remediation residual saturation levels. As shown in Table 1, the saturated thickness in the cell was different for the pre- and post-PTT's. The increased saturated thickness for the post-PTT caused the tracer solution to flow through zones not treated by the surfactant solution and not "seen" during the pre-PTT. Adjusting the residual saturation estimates for this "new" area using the soil core data improved the validity of the results. The mean travel times and residual saturation values for the tracers are shown in Table 2. The average post-remediation residual value is 2.71%, compared to the average pre-remediation residual value of 8.52%. This represents a 68% reduction in the residual saturation, which is higher than that derived from the pre- and post-demonstration soil cores (58%).

In order to determine mass recovery for the surfactant, both injection and

54

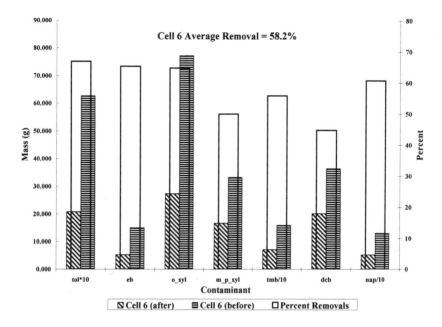

Figure 2. Treatment zone soil concentrations and removal percentages for Cell 6.

Table 2: Cell 6 PTT Results

| Extraction Well | Mean Travel Time (minutes) | | | | | | Residual Saturation (%) | |
| | Bromide | | 2,2-dimethyl-3-pentanol | | Hexanol | 6-methyl-2-heptanol | | |
	Pre	Post	Pre	Post	Pre	Post	Pre	Post
1	3009	1639	7218	3573	1104	4360	9.71	3.29
2	1730	2009	3926	3988	1164	5327	8.90	3.27
3	1570	2088	3072	3433	1041	3703	6.96	1.56
						Average	8.52	2.71
						% Removal	68.24	

[a] Values calculated with using total saturated thickness.
[b] Values calculated with excess saturated zone removed.

extraction well samples were analyzed. Influent mass was calculated based on 60 injection well samples (20 samples from each of the three injection wells). The total mass recovered was calculated based on 129 samples (43 samples from each of the three extraction wells). The percent recovery of Dowfax 8390 integrated over the three extraction wells was 93%. Surfactant mass recovered in the third extraction well was low because water was injected in the fourth injection well on the same side of the cell, i.e., surfactant was only injected in the first three injection wells.

The mass of contaminants in the effluent samples can be represented by Total Petroleum Hydrocarbons (TPH), which is defined as the sum of all LNAPL constituents in solution. Changes in the concentrations of TPH and surfactant for the first and third extraction wells are plotted in Figures 3 and 4. For each of the extraction wells, there is a noticeable initial hump in both the TPH and surfactant curves, followed by a plateau region up until the injection solution is switched to water. The initial humps in the curves are due to the elevated concentrations of surfactant in the injection solution due to incomplete mixing in the holding tank. Comparing the concentration levels between the figures shows the effects of dilution in the third extraction well. Table 3 is a summary of the estimated amount of contaminants removed at each extraction well.

Thus, as anticipated, solubilization with Dowfax 8390 was a very robust, user-friendly system. While more efficient solubilization systems exist, these systems are chemically more complex and inherently more difficult to deal with. The removal efficiency via solubilization was significantly greater than by water alone (< 1% in the same number of pore volumes).

Surfactant-Enhanced Mobilization

Initial laboratory studies to identify middle phase compositions were conducted using JP-4, the major fraction of the NAPL at Hill AFB. However, compositions established with the JP-4 were not successful in achieving middle phase systems with the actual Hill NAPL, indicating that significant weathering of the contaminant had occurred. Modification of these surfactant/cosurfactant systems achieved middle phase systems with the Hill NAPL.

The surfactant mobilization demonstration study was preceded by the pre-PTT and a pre-flush of two pore volumes of water. During the demonstration study, the surfactant/cosurfactant solution significantly reduced interfacial tension, thus mobilizing the NAPL. While samples did evidence excess oil, they did not evidence the occurrence of three phases (i.e., water, oil, and middle phase). It could be that the optimal conditions developed in the laboratory were not exact for the heterogeneous field conditions. In any event, excellent contaminant elution was realized even in the absence of the ideal middle phase. This is consistent with the continual decreases in IFT and increase in solubilization potential as the middle phase region is approached

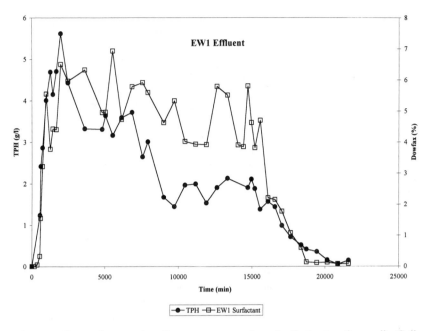

Figure 3. Contaminant and surfactant concentrations for first extraction well - Cell 6.

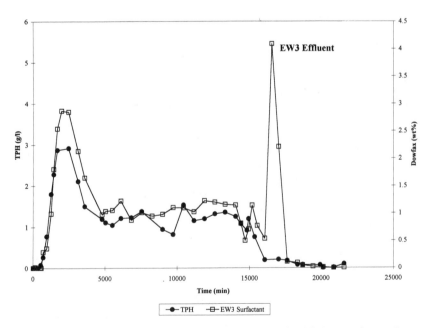

Figure 4. Contaminant and surfactant concentrations for third extraction well - Cell 6.

Table 3: Cell 6 Contaminant Removal Estimated from GC Analysis

Location	Effluent TPH Measured (kg)	Contaminant Mass Removed* (g)
EW 1	35.7	622
EW 2	35.9	598
EW 3	17.4	291

*Sum of target analytes

(see surfactant fundamentals chapter), and demonstrates that this approach is more robust than may initially be perceived. The surfactant mobilized NAPL solution had a higher viscosity than the aqueous solutions previously flushed through the cell (water or partitioning tracers). The increased viscosity of the fluid resulted in a decrease in hydraulic conductivity. Because the injection flow rate was held constant, the fluid level in the injection wells began to rise, rising by as much as ten to fifteen feet. When the injection fluid was switched back to water, the fluid levels receded back to their pre-flush values.

Figure 5 is a plot of the treatment zone contaminant concentrations and removal percentages for Cell 5. The removal percentages range from 88% (DCB) to greater than 99% (toluene, ethylbenzene, o-xylene, m-xylene) with an arithmetic average removal of 97.2%

A qualitative (visual) analysis of the pre-mobilization PTT data revealed that the hexanol response curves showed breakthrough earlier than bromide and the tails dissipated slightly earlier, as observed in Cell 6 (and attributed to biodegradation). This is not consistent with the expected behavior given the partition coefficients of the two chemicals. Therefore, the hexanol data was considered unreliable and was not used in the residual saturation calculations.

The post-mobilization PTT breakthrough curves appeared to show significantly less separation than the respective pre-mobilization curves. However, moment calculations using bromide and 2,2-dimethyl-3-pentanol indicated that residual saturation levels had actually increased, obviously at odds with visual observations and quantitative measurements. This apparent increase can be explained by the extremely long tailing effect associated with the two alcohol tracers. It was hypothesized that the alcohols were partitioning into and between residual surfactant micelles. The concentration of the Tween 80 remaining in the cell ranged up to 0.5 wt%, which is much higher than its CMC. Laboratory experiments to evaluate alcohol partitioning with residual Tween 80 are planned for further validation.

The tailing effect in the PTT breakthrough curves was addressed by comparing the two alcohols (since their tails would both be similarly impacted). Using this approach, the residual saturations were shown to have decreased. The mean travel times and residual saturation values are shown in Table 4. The average post-remediation residual saturation value is 2.52%, compared to the average pre-remediation value of 4.31%. This represents only a 42% reduction in residual saturation levels which is significantly lower than that derived from the pre- and post-demonstration soil cores.

Influent surfactant mass was calculated based on 16 injection well samples. The total mass recovered was calculated based on 51 extraction well samples (17 samples from each of the three extraction wells). The percent recoveries for all three extraction wells were: 119% for Aerosol OT (AOT) and 109 % for Tween 80, obviously indicating errors in measuring surfactant concentrations and/or flow rates,

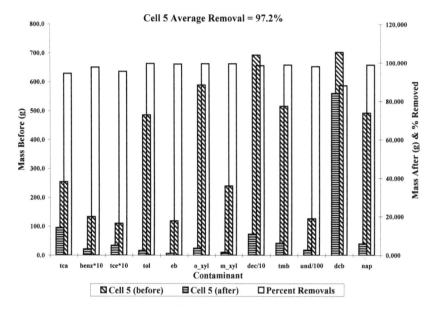

Figure 5. Treatment zone soil concentrations and removal percentages for Cell 5.

Table 4: Cell 5 PTT Results

| Extraction Well | Mean Travel Time (minutes) | | | | | | Residual Saturation (%) | |
| | Bromide | | 2,2-dimethyl-3-pentanol | | Hexanol | 6-methyl-2-heptanol | | |
	Pre	Post	Pre	Post	Pre	Post	Pre	Post
1	1267	591	1826	1333	1074	2114	3.28	2.04
2	1141	525	1659	1614	1124	2447	3.37	1.74
3	1214	614	2270	1267	1625	2444	6.27	3.77
						Average	4.31	2.52
						% Removal	41.56	

but demonstrating that a vast majority of the surfactant was recovered. Changes in concentrations of TPH and the surfactants for the middle extraction well are plotted in Figure 6. Mobilization of the trapped NAPL is evident from these figures as the elevated concentrations of TPH precede breakthrough of the surfactants. Table 5 is a summary of the estimated amounts of contaminants removed at each extraction well. Again, mobilization greatly expedited contaminant removal over water alone (80 to 90% versus < 1%, respectively).

Comparison

The contaminant removals for both technologies are shown in Table 6. In general, the changes in residual saturation determined from the PTT's and the TPH mass extracted both reflect the relative removals evidenced in the soil cores. The superior removal efficiency of the mobilization mechanism is reflected in the magnitude of the TPH extracted from the cells. Although hampered by operational limitations for Cell 6, the Dowfax 8390 solution showed steady removal of NAPL constituents from those zones flushed with supra-micellar solution. Total contaminant mass removed from the cell could have been much higher if more pore volumes of surfactant solution were flushed through the cell or if higher surfactant concentrations had been used.

The two systems can also be compared by analyzing the extraction well data. The mobilization data (Figure 6) has a higher peak concentration and doesn't evidence the plateau evident in the solubilization data (Figure 3). The plateau in solubilization data illustrates that interface mass transfer that is occurring. Although at a much higher rate than for water alone; the contaminants are still dissolving out of the trapped oil. The absence of this plateau for the mobilization data illustrates the bulk displacement evidenced with this system. Obviously this is advantageous in terms of both efficiency of removal and extent of remediation.

The PTT results show tremendous variation in residual saturation levels, both between cells and within each cell. In addition, the conservative and partitioning tracers performed inconsistently between the pre- and post-demonstration PTTs in both cells. As a result, residual saturations were calculated using different combinations of conservative and partitioning tracers for the two PTTs. Poor mass recoveries for the alcohols during the post-demonstration PTTs are attributed to biodegradation, but this has not been confirmed. All of these issues raise concerns as to the validity of the PTT results obtained in the study.

Thus, mobilization proved to be much more efficient than solubilization, as expected. The mobilization system was more sophisticated to achieve, and difficult to implement, illustrating the potentially more robust nature of the solubilization system. Further development of middle phase systems should improve their robustness, thereby further enhancing the attractiveness of this approach.

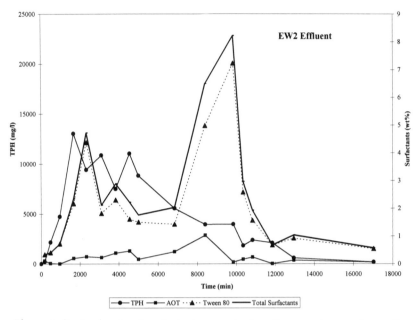

Figure 6. Contaminant and surfactant concentrations for second extraction well - Cell 5.

Table 5: Cell 5 Contaminant Removal Estimated from GC Analysis

Location	Effluent TPH Measured (kg)	Contaminant Mass Removed* (g)
EW 1	73	3374
EW 2	78	3590
EW 3	83	10809

*Sum of target analytes

Table 6: Comparison of Contaminant Removals

Cell	Contaminant Removed	Locations			
		EW 1	EW2	EW 3	Soil Cores
6	TPH Extracted (kg)	36	36	17	
	Mass of Target Analytes Removed (g)	622	598	291	265
5	TPH Extracted (kg)	73	78	83	
	Mass of Target Analytes Removed (g)	3374	3590	10809	22459

Summary

Both the solubilization and mobilization mechanisms significantly elevated the extraction of the majority of the NAPL constituents in each demonstration cell versus water alone. The mobilization mechanism was more effective than the solubilization mechanism in terms of contaminant removal for approximately the same amount of contaminant mass; however, the binary surfactant system used for mobilization was more difficult to work with and caused a significant increase in the viscosity of fluids in the subsurface which led to pronounced mounding of the fluids. The Dowfax solution was extremely easy to work with, hence, additional flushing would be relatively straightforward. These results show the significant potential of surfactant systems to expedite pump-and-treat remediation of residual oil, and encourage the continued development and implementation of these systems.

Acknowledgments

This study was funded by the Strategic Environmental Research and Development Program (SERDP), Project Number CU 368. Supplemental financial and technical support was provided by the Environmental Restoration and Management Division, Hill AFB USAF.

Literature Cited

1.   Olsen, R.L.; Kavanaugh, M.C. *Water Environment & Technology,* **1993**, 42-47

2.   Mackay, D.M.;Cherry, J.A. *Env. Sci. & Tech.*, **1989**, *23*, 630-636.

3.   Haley, J.L.; Hanson, B.; Enfield, C.; Glass, J. *Ground Water Monitoring Review,* **Winter 1991**, 119-124.

4.   Schmelling, S.G.; Keely, J.W.; Enfield, C.G. In *Contaminated Land Treatment Technologies*, Rees, J.F., Ed.; Elsevier Applied Science, NY, 1992, 220-233.

5.   National Research Council (NRC) *Alternatives for Ground Water Cleanup*; National Academy Press, Washington, DC ,1994.

6.   Palmer, C. D. and Fish, W., *Chemical Enhancements to Pump and Treat Remediation*, EPA/540/S-92/001, USEPA, Ada, OK, 1992.

7.   *Surfactant Enhanced Subsurface Remediation: Emerging Technologies*; Sabatini, D. A.; Knox, R. C.; Harwell, J. H., Eds.; ACS Symposium Series 594, American Chemical Society, Washington, D.C., 1995.

8.   Shiau, B. J.; Sabatini, D. A.; Harwell, J. H.;Vu, D. *Env. Sci. & Tech*, **1996**, *30*, 97-103.

9.   Jin, M.; Delshad, M.; Dwarakanath, V.; McKinney, D.C.; Pope, G.A.; Sepehrnoori, K.; Tilburg, C.; Jackson, R.E. *Water Resources Research*, **1995**, *31*, 1201-1211.

10.   Knox, R.C.; Sabatini, D.A.; Harwell, J.H.; Brown, R.E.; West, C.C.; Blaha, F.; Griffin, C. *Ground Water*, **1997**, *35*, 948-953.

Chapter 6

Demonstration of Surfactant Flooding of an Alluvial Aquifer Contaminated with Dense Nonaqueous Phase Liquid

C. L. Brown[1], M. Delshad[2], V. Dwarakanath[2], R. E. Jackson[3],
J. T. Londergan[3], H. W. Meinardus[3], D. C. McKinney[1], T. Oolman[4],
G. A. Pope[2], and W. H. Wade[5]

Departments of [1]Civil Engineering, [2]Petroleum and Geosystems Engineering,
and [5]Chemistry, The University of Texas at Austin, Austin, TX 78712
[3]Duke Engineering and Services, Inc., 911 Research Boulevard,
Austin, TX 78758
[4]Radian International LLC, P.O. Box 201088, Austin, TX 78720

Two field tests at Hill Air Force Base Operational Unit 2 were completed in May and September of 1996 to demonstrate surfactant remediation of an alluvial aquifer contaminated with DNAPL (dense nonaqueous phase liquid). The DNAPL at Hill OU2 consists primarily of trichloroethene (TCE), 1,1,1-trichloroethane (1,1,1-TCA) and tetrachloroethene (PCE). Sheet piling or other artificial barriers were not installed to isolate the 6.1 x 5.4 m test area from the surrounding aquifer. Hydraulic confinement was achieved by: (1) injecting water into a hydraulic control well south of the surfactant injectors (2) designing the well pattern to take advantage of the alluvial channel confined below and to the east and west by a thick clay aquiclude and (3) extracting at a rate higher than the injection rate within the well pattern. An extensive program of laboratory experimentation, hydrogeological characterization, effluent treatment and predictive modeling was critical in the design of these tests and the success of the project. Simulations were conducted to determine test design variables such as well rates, injected chemical amounts and test duration, and to predict the recovery of contaminants and injected chemicals, degree of hydraulic confinement and pore volume of the aquifer swept by the injected fluids. Partitioning interwell tracer tests were used to estimate the volume and saturation of DNAPL in the swept volume and to assess the performance of the surfactant remediation. Analysis of the Phase I and Phase II results showed high recoveries of all injected chemicals, indicating that hydraulic confinement was achieved without sheet pile boundaries. Approximately 99% of the DNAPL within the swept volume was removed by the surfactant in less than two weeks, leaving a residual DNAPL saturation of about 0.0003. The concentration of dissolved contaminants was reduced from 1100 mg/l to 8 mg/l in the central monitoring well during the same time period.

The conventional method of treating DNAPL-contaminated aquifers to-date is 'pump and treat', where contaminant dissolved in groundwater and, possibly, DNAPL itself are pumped to the surface and treated. This method can be quite effective at removing the more mobile DNAPL within the drainage area of the pumping wells and also at minimizing the migration offsite of the contaminated groundwater plume. Unfortunately, conventional pump and treat methods have proved totally ineffective at removing the DNAPL saturation remaining as less-mobile, more isolated ganglia within the groundwater aquifer, sometimes referred to as trapped, bypassed, or residual DNAPL *(1)*.

Surfactants have recently shown great promise in remediating this trapped DNAPL, in laboratory experiments *(2-4)*, small-scale field tests *(5-7)*, and aquifer simulation studies *(8-9)*. The addition of surfactants to water injected into aquifers has the potential to greatly enhance the remediation efficiency by: (1) increasing the solubility of the solvent contaminants in groundwater up to several orders of magnitude ('solubilization') and (2) by decreasing the interfacial tension between the DNAPL and the water, thereby reducing the capillary forces 'trapping' the DNAPL in the pore spaces and making the residual DNAPL more mobile ('mobilization'). Which one of these two processes dominate, or indeed which is most desirable, is a function of the characteristics of the site, the contaminant, and the surfactant.

Partitioning tracer tests are conducted before surfactant remediation to estimate the volume of DNAPL in the swept volume of the demonstration area and again after remediation for performance assessment of the remediation. Partitioning tracer tests have been used for this purpose at several sites recently. These tests include the saturated zone tests at Hill Air Force Base Operable Unit 1 *(11)* and at a Superfund site in Arizona *(12)* and an unsaturated zone test in the Chemical Landfill Waste site near Sandia National Laboratory *(13)*.

Among the most important new achievements of this field demonstration of surfactant flooding were the following:

(1) Demonstrated that surfactant flooding can remove almost all of the residual DNAPL from the swept volume of an alluvial aquifer in a very short time period. In less than two weeks, 99% of the DNAPL was removed from the volume swept by the surfactant. The final DNAPL saturation was 0.0003, which corresponds to 67 mg/ kg of soil. The goal was to remove the source of the contaminant plume rather than the contaminants dissolved in the water; however, the dissolved contaminants were reduced from 1100 mg/l to 8 mg/l at the central monitoring well and were still declining when the pumping was stopped.

(2) Demonstrated the use and value of partitioning tracer tests before and after the remediation to determine the amount of DNAPL present before and after remediation. This was the first such test at a DNAPL site.

(3) Demonstrated the use and value of predictive modeling to design the test and to address issues critical to gaining approval for a surfactant flood of a DNAPL source zone in an unconfined aquifer, such as hydraulic confinement, injected chemical recovery, DNAPL recovery, and final concentrations of injected chemicals and contaminants.

This paper focuses on the analysis and simulations needed to design surfactant remediation field tests and briefly summarizes the key results of the Phase I and Phase II field tests.

Spent degreaser solvents and other chemicals were disposed of in shallow trenches at the Hill Air Force Base Operational Unit 2 (Hill AFB OU2) Site, located north of Salt Lake City, Utah, from 1967 to 1975. These disposal trenches allowed DNAPL to drain into an alluvial aquifer confined on its sides and below by thick clay, resulting in the formation of a DNAPL pool. The DNAPL at Hill OU2 consists primarily of three chlorinated solvents or VOCs (volatile organic compounds). These are trichloroethene (TCE), 1,1,1-trichloroethane (1,1,1-TCA), and tetrachloroethene (PCE).

In 1993, Radian installed a Source Recovery System (SRS) consisting of extraction wells and a treatment plant. More than 87,000 L of DNAPL have been recovered and treated by this system *(14)*. This still leaves much DNAPL source remaining at the site in the form of residual or bypassed DNAPL not recovered by pump and treat operations that will continue to contaminate the groundwater.

The primary objectives for the Phase I test were to: (1) determine the amount of DNAPL initially present using partitioning tracers (2) achieve and demonstrate hydraulic control of the surfactant solution at the site (3) test both the surface treatment and subsurface injection-extraction facilities (4) obtain data for the design of the Phase II remediation test and for regulatory purposes (5) measure the swept volume of each well pair using tracers (6) validate tracer selection and performance (7) measure the hydraulic conductivity during injection of surfactant solution in one well and (8) optimize the sampling and analysis procedures.

The primary objectives for the Phase II test were to: (1) use commercially available biodegradable chemicals to remove essentially all DNAPL in the swept volume of the well pattern (2) recover a high percentage of all injected chemicals and leave only very low concentrations of these chemicals in the groundwater at the end of the test (3) use partitioning tracers to accurately assess remediation performance (4) maintain hydraulic control (5) use existing surface treatment facilities on site to treat the effluent during the test and (6) complete the entire test including before and after tracer tests within 30 days.

The Phase II test of August, 1996, lasted approximately one month and consisted of the initial water flood, a pulse of tracer injection, water flooding (NaCl was added to the water during the final day of this flood), the surfactant flood, water flooding, a second pulse of tracer injection, water flooding and finally a short period of extraction only.

Approximately one hundred simulation predictions were conducted to design tests that would achieve the Phase I and Phase II objectives in the shortest time and within budget. These simulations used an aquifer model based on Hill AFB site characterization including field hydraulic testing and well data, and extensive laboratory experiments using Hill AFB soil, DNAPL, groundwater and injected tap water.

Design of the Field Tests

The Hill AFB OU2 site characterization included the following: aquifer stratigraphy and aquiclude topography; porosity and permeability distribution; soil, groundwater, and contaminant constituents and distribution; hydraulic gradient direction, magnitude and seasonal variation; aquifer temperature and seasonal variation. This site characterization was based on the following site data: soil borings, well logs, seismic data, water levels, soil contaminant measurements, DNAPL and groundwater sampling and analysis, hydraulic testing, and historical pumping data.

Site Description and Characterization. Figure 1 shows the OU2 Site at Hill AFB and the locations of the test area wells and nearby wells. Within the test area, there are a line of extraction wells (U2-1, SB-1, SB-5) to the north 3.1 m apart and a line of injection wells (SB-3, SB-2, SB-4) 3.1 m apart and located 5.4 m south of the line of extraction wells. This 6.1 m x 5.4 m approximately square test area well configuration is also referred to as a 3x3 line drive pattern. A monitoring well, SB-6, is located in the center of the test area. Additional monitoring wells (for fluid levels and water samples) are located to the north and south of the mapped area. The site's abandoned chemical disposal trenches, used for disposal of spent degreasing solvents, are located to the south of U2-1; the exact location is unknown.

A hydraulic control well, SB-8, was located 6.1 m south of the line of injection wells. The injection wells within the test area inject water and various chemicals while the hydraulic control well is located outside of the test area and injects water only. The purpose of the hydraulic control well is to prevent the migration of injected chemicals to the south of the test area. The pattern is confined by the aquiclude to the east and west, by extraction wells to the north, and by the hydraulic control well to the south. The choice of appropriate locations and rates of the seven wells are critical in achieving this confinement and are key design parameters. More than one hydraulic control well would likely be needed in most surfactant floods, but in this case one was sufficient due to the favorable channel geometry of the aquiclude. The injection and extraction rates are high enough that the forced gradient completely dominates the hydraulic gradient between wells during the test. This is an essential part of a successful surfactant flood.

The depth to the water table, approximately 1423 m above mean sea level (AMSL), is 6 to 8 m below ground surface in the U2-1 area, and varies seasonally. The depth to the Alpine clay underlying the aquifer is contoured on Figure 1 to this same depth of 1423 m AMSL. The Alpine Formation is on the order of a hundred meters thick and bounds the aquifer below and to the east and west and forms a very effective aquiclude for the aquifer. The aquifer is in a narrow channel with a north to south trend. A more complete site description may be found in *(15-17)*. From October 1993 to June 1994, 87,000 L of DNAPL and over 3,800,000 L of contaminated groundwater were produced from these areas *(14)*.

Groundwater flow is towards the northeast, and varies in direction and magnitude seasonally. In the test area, the hydraulic gradient is around 0.002 *(17)*. This natural hydraulic gradient is approximately two orders of magnitude less than the forced gradients induced during the field tests.

68

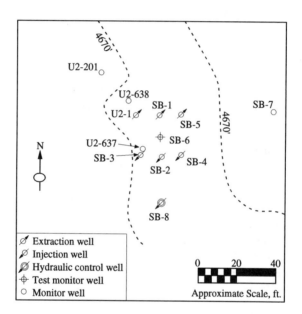

Figure 1. Plan view of Hill AFB OU2 site.

Many pumping tests have been conducted in the OU2 area over the past eight years, with resulting hydraulic conductivities ranging from 3.5×10^{-5} to 4.1×10^{-4} m/s, and, assuming only water is present in the zone, equivalent to a permeability of 3.6 to 44 μm^2 (3.6 to 44 Darcy) *(17)*. In Oct. 1996, a series of pump tests were conducted for wells in the test area, yielding hydraulic conductivities ranging from 9.5×10^{-6} to 1.4×10^{-4} m/s, equivalent to a permeability of 1 to 14 μm^2 *(16)*. Because the soil is unconsolidated, obtaining representative in-situ permeability measurements from the cores is difficult but column values of hydraulic conductivity are on the order of 10 μm^2.

Contaminant Characterization. It is very important to know the volume and distribution of DNAPL before remediation is started, yet this is usually very poorly known. The purpose of a partitioning tracer test is estimate the volume and saturation of DNAPL throughout the test volume and provides a spatially integrated value with a minimum of disturbance of the soil or DNAPL. Some estimate of the DNAPL volume was needed for the Phase I tracer test design simulations. This initial DNAPL saturation distribution was estimated based upon: (1) soil contaminant concentrations measured from soil samples collected when the wells were drilled (2) aquiclude structure, (3) measured DNAPL volumes produced from some wells (4) and produced contaminant concentration history from extraction wells within and outside the test pattern. Although the uncertainty using these data is high, it turned out to be a sufficiently good estimate of DNAPL volume for tracer test design purposes.

Contaminant measurements in the soil samples acquired before any production from the test area showed DNAPL in the lower two meters of a narrow channel filled with sand and gravel. For the Phase I design simulations, the DNAPL saturation was approximated as 0.20 in the bottom three layers of the six-layer aquifer simulation model (excluding aquiclude regions), representing the lowest 2 meters of the aquifer. The upper 4 meters of the aquifer (the upper three model layers) were assumed to contain no DNAPL in the test area. By volumetrically averaging the initial saturations throughout the test volume, the initial aquifer DNAPL saturation was estimated to be approximately 0.03.

Surfactant Phase Behavior. Extensive laboratory experiments were conducted to establish an effective surfactant formulation. This involved batch phase behavior tests, measurements of viscosity, interfacial tension, tracer partition coefficients and numerous tracer and surfactant column floods (4,19). These experiments used soil, DNAPL, groundwater, and tap water from the site. The phase behavior experiments were used to identify and characterize suitable surfactants that form classical microemulsions and to identify the need for co-solvent to eliminate problems with liquid crystals, gels or emulsions, which can cause soil plugging. Co-solvent also promotes rapid equilibration and coalescence to the desired equilibrium microemulsions. The phase behavior of the surfactant was measured as a function of electrolyte concentrations, temperature, co-solvent concentration and other key variables. The soil column experiments were used to evaluate the tracers, to assess the effectiveness of surfactants at removing DNAPL from the soil, to measure surfactant adsorption on the soil, to assess any problems with reduction in hydraulic conductivity and to evaluate the use of co-solvents in improving the test performance.

The anionic surfactant used in these tests was sodium dihexyl sulfosuccinate obtained from CYTEC as Aerosol MA-80I. Extensive testing of this surfactant was done and the results can be found in (4,18, 19). The solubility of the Hill DNAPL in a microemulsion containing 8% dihexyl sulfosuccinate and 4% co-solvent (isopropyl alcohol, IPA) was determined as a function of NaCl added to the Hill tap water. The solubility of the three principal Hill chlorinated DNAPL constituents in groundwater is about 1,100 mg/L. Adding 7000 mg/L NaCl to the mixture at 12.2° C increases the contaminant solubility to approximately 620,000 mg/L of microemulsion, or a solubility 560 times greater than that in groundwater.

Partitioning Tracer Experiments.   Seven tracers were selected for use as conservative and partitioning tracers at the Hill OU2 site (19). In order to use the partitioning tracer tests to estimate DNAPL saturations, it is essential to know the partition coefficients of the tracers between DNAPL and water accurately. Batch equilibrium partition coefficient tests were performed to measure the partition coefficients of the alcohol tracers. The partition coefficient is the ratio between the concentration of the tracer species in the DNAPL and the concentration of the tracer species in the aqueous phase. The optimum injection rates, in terms of desired retention times for both surfactant and tracer tests, were determined through column experiments. These experiments indicated that a retention time greater than about 20 hours is needed to achieve local equilibrium, essential for obtaining good estimation of residual DNAPL saturations using partitioning tracer tests, and this same constraint was assumed to apply to the field test. Partitioning tracers for estimation of DNAPL contamination is described in detail by Dwarakanath (*19*), Jin *et al. (20),* Jin *(21),* and Pope *et al. (22).*

Surfactant Column Experiments.   Extensive surfactant flood experiments were conducted using columns packed with soil from the OU2 site test area, as well as DNAPL from the site. The surfactant mixture used in the final design had been shown by these column experiments to reduce the DNAPL saturation in the soil to less than 0.001 as estimated from both the partitioning tracers and mass balance. Although our goal was to remove the DNAPL rather than the dissolved contaminant in this unconfined aquifer system, laboratory column data showed that the TCE concentration in the effluent water could be reduced to less than 1 mg/L after surfactant flooding of the soil. The concentration of TCE and other VOCs and tracers was measured using a Gas Chromatograph (GC) with an FID mechanism with straight liquid injection. The minimum detection of TCE concentration was 1 mg/L.

Surfactant adsorption was measured in column experiments using Hill soil by comparing the response of a conservative tracer (tritiated water) to that of the surfactant labeled with carbon 14. The retardation factor for the surfactant was 1.00094. The surfactant adsorption calculated from this value is 0.16 mg/g of soil, which is zero within experimental error of the retardation factor. All of these and other experimental results are described and discussed in detail in (4,*19*).

Surface Treatment. Existing groundwater treatment facilities at Hill AFB OU2, including phase separators and a steam stripper *(14)*, were utilized to treat all of the groundwater and DNAPL recovered during the Phase I and II field tests. The high levels of surfactant, co-solvent and contaminant in the recovered groundwater presented significant challenges for steam stripper operation. Prior to the field tests, the ASPEN model was used to model the treatment facilities to determine if and how the existing steam stripper could achieve required contaminant removal levels. The predicted composition of the effluent from the UTCHEM modeling described below was used as the input to ASPEN. Actual operation of the steam stripper during the field tests demonstrated that predicted performance levels could be achieved. It was also shown that steam stripping is a favorable technology for the treatment of the highly contaminated groundwaters recovered during surfactant enhanced remediations. These results can be found in the final report to the Air Force Center for Environmental Excellence (24).

Aquifer Model Development and Simulations. UTCHEM, the University of Texas Chemical flood simulator, was used to design the tests and to predict the performance of the Phase I and Phase II surfactant and tracer tests. UTCHEM is a three-dimensional, multicomponent, multiphase (water, NAPL, microemulsion, air, soil) compositional simulator incorporating higher-order finite-difference methods with a flux limiter. Use of this simulator makes possible the study of phenomena critical to surfactant flooding such as solubilization, mobilization, surfactant adsorption, interfacial tension, capillary desaturation, dispersion/diffusion, and the microemulsion phase behavior. A detailed discussion of the formulation and features of UTCHEM may be found in Delshad et al. *(25)*. The use of the UTCHEM simulator in modeling DNAPL contamination and remediation processes is discussed in Brown et al. *(9)* and more detail on the specific modeling of the Phase I and Phase II field tests can be found in (26).

Determining realistic in-situ properties for these unconsolidated soil samples is very difficult, due to grain rearrangement and disruption of the porous media during boring, transport, cutting, and storage. The difficulty is increased for the OU2 site soil samples because the aquifer is composed primarily of gravel interspersed with cobbles, some of them larger than the 6 cm diameter sampling tube. To minimize the core disruption, the sampling tube was frozen upon reaching the laboratory, before the core cutting and until the measurements could be obtained.

The model grid and aquifer properties are summarized in Table I. The steeply dipping lower boundary of the aquifer was modeled by assigning lower permeability (5×10^{-6} μm^2) and porosity to all gridblocks lying within the aquiclude. This aquiclude structure, and the ability to model it accurately, played a key role in the design of the field test, and combined with the use of a hydraulic control water injection well to the south of the surfactant injection wells, allowed hydraulic control to be achieved without using sheet piling. The sandy/gravelly aquifer soil was modeled using a random correlated permeability field with a standard deviation of ln k of 1.2. The correlation length along the channel was 3 meters, across the channel 1.5 meters and in the vertical direction 0.3 meters. The permeability assigned to individual gridblocks

Table I: Grid and Aquifer Properties Used in the Phase I Design Simulations

Property	Value	References and Comments
Mesh	xyz: 20 x 17 x 6 (2040 gridblocks)	
Dimensions	54 x 20 x 5.9 meters	
Mesh size	0.8 x 0.8 x 0.5 meters, (smallest aquifer cell), pore volume of 72 liters 14 x 2.7 x 2.0 meters, (largest aquifer cell), pore volume of 20,100 liters	
Boundary conditions	Impervious top, bottom, east and west boundaries; constant potential boundaries north and south	
Initial pressure	Atmospheric pressure in top layer; hydrostatic distribution in vertical	
Initial DNAPL saturation	20% (the DNAPL residual saturation) in the lower 2 m of the aquifer	Based on core contaminant measurements and measured DNAPL pool depth
Aquifer pore volume	126,000 liters	
Total aquifer DNAPL volume	5090 liters	Including DNAPL in the northern primary DNAPL pool and other DNAPL outside test area

ranged from a low of 0.2 μm^2 to a high of 420 μm^2. These values resulted in a good match of the tracer data taken with the same well field during Phase I.

The natural hydraulic gradient (0.002) is approximately two orders of magnitude less than the induced gradient during the field tests (0.10-0.20), so the natural gradient was assumed to be zero for the simulations. Open boundaries were placed at the north and south sections of aquifer model to allow flow into and out of the aquifer in response to the test area injection and extraction.

In UTCHEM, phase behavior parameters define the solubility of the organic contaminant in the microemulsion as a function of surfactant, co-solvent and electrolyte concentrations and temperature. These parameters were obtained by matching the experimentally determined solubility of the Hill contaminants at various surfactant, co-solvent and electrolyte concentrations. The phase behavior model agrees well with the measured data. Experimentally determined interfacial tensions for DNAPL-groundwater and DNAPL-microemulsion were used to calibrate the UTCHEM correlation for calculating interfacial tension.

Many UTCHEM simulation cases were performed to determine design parameters such as hydraulic control well injection rate, injection and extraction rates, frequency of sampling points, amount of surfactant, composition of injected surfactant solution, amount and composition of tracer solutions, duration of water flooding and extraction needed after the surfactant injection, and the concentrations of contaminants, surfactant and alcohol in the effluent. The high injection rate of water in the hydraulic control well to the south of surfactant injection wells was found to a particularly important design variable.

Results and Discussion

Phase I Field Test. The Phase I test, completed in May 1996, lasted approximately two weeks and consisted of an initial water flood, tracer injection, water flooding, injection of a small mass of surfactant in one well only, water flooding, then post-test extraction to recover any remaining injected chemicals. Water flooding consists of injection of water only to sweep the fluids within the test area volume towards the extraction wells where they are pumped and treated at the surface.

Based upon tracer concentrations measured during the test with on-site GCs, about 97% of the tracers injected during the two weeks of the Phase I Field Test were recovered. This high tracer recovery was due to good hydraulic confinement of the test area. This is the primary confirmation and best means of determining the degree of hydraulic control; however, the evaluation that hydraulic control was achieved in the Phase I tests can also be supported by three other sources of information: Firstly, measured piezometric data during the tests indicated that water levels for the three extraction wells were approximately 0.5 meters lower than the surrounding aquifer, creating a large gradient from within the test area towards the extraction wells. Secondly, monitoring wells had very low measured concentrations (at or below the measurement detection level) of the injected tracers throughout the test. These monitoring wells were placed both to the north and south (aquiclude confines aquifer to the east and west) approximately 21 meters away from the test area. Finally, simulation results matched the model very well in predicted tracer recovery (both were 97%).

Figure 2 shows Phase I tracer concentrations for the central extraction well SB-1. Four alcohol tracers were injected during the Phase I test, with partition coefficients ranging from 0 to 141. The upper graph compares the predicted tracer concentrations with the field data measured during the Phase I test, for the two tracers used in the moment analysis to estimate the initial DNAPL saturations and volumes. The lower graph in Figure 2 shows the produced concentrations for all four of the injected tracers, plotted on a log scale to highlight the log-linear behavior typically exhibited during latter part of the tracer tests. There is substantial separation between the nonpartitioning tracer, isopropanol, and the highest partitioning tracer, 1-heptanol, as the latter is retarded by the presence of the DNAPL.

The predicted concentrations shown here are those published in the workplan before the test was undertaken and therefore are not history-matched or calibrated to the field data. Even so, breakthrough times, peaks, and tails for both the partitioning and nonpartitioning tracers are similar to the UTCHEM predictions. Because the simulation predictions agreed with the actual Phase I field performance very well, few modifications were required in the aquifer model for the Phase II design simulations. Approximately 750 L of contaminant was extracted during the Phase I test based upon the partitioning tracer data.

Phase II Field Test. Phase II included an initial tracer test, a NaCl preflood, a 2.4 PV surfactant flood, and a final tracer test, followed by a period of extraction only to maximize the recovery of injected chemicals. See Table II for a summary of the Phase II test. The purpose of the one day NaCl preflood was to increase the salinity of the water to a value closer to the optimal value of 7000 mg/l NaCl before the surfactant was injected. The swept pore volume calculated from the non-partitioning tracer analysis is approximately 57,000 liters for all three well pairs and the injected pore volumes listed in Table 1 are based on this total swept pore volume. The saturated pore volume of the alluvium within the line drive well pattern as estimated from the structure of the clay aquiclude below the aquifer was also about 57,000 liters, indicating very little if any unswept soil between the screened intervals of the injection and extraction wells.

Figure 3 shows the measured surfactant concentration for the central extraction well SB-1 during Phase II and compares those with UTCHEM predictions. While the breakthrough and peak times are similar, the magnitude of the peak and the 'tail' concentrations are significantly different. The surfactant concentration dropped below the CMC (critical micelle concentration) at around 13 days in the UTCHEM prediction case and around 18 days in the field test. These differences in observed and predicted surfactant concentrations are due in part to differences in the design rates and those actually achieved in the Phase II field test. The lower extraction rates in the field test result in lower extraction/injection ratios, less dilution from groundwater flowing into the extraction wells from the north and higher surfactant concentrations than the model prediction shown in Figure 3. There is an increase in surfactant concentration at 20 days due to the increase in extraction rates at the end of the water flood. These and other factors could be adjusted and would improve the

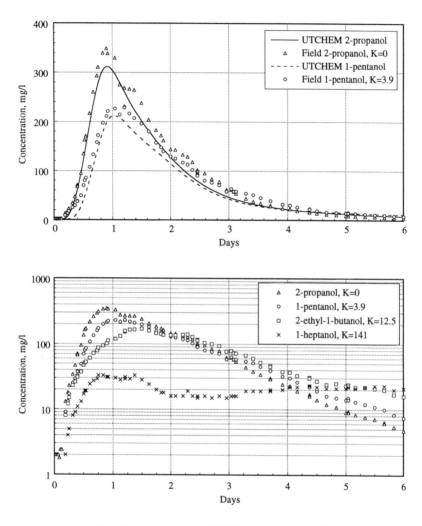

Figure 2. Tracer data for extraction well SB-1. Top: comparison of UTCHEM prediction with 2-propanol and 1-pentanol. Bottom: tracer data plotted on a log scale.

Table II: Summary of Phase II Test

Duration (days)	Cumulative time (days)	Segment	Pore Volumes	Chemicals added to Hill source water
1.7	1.7	water flooding	1.2	
0.4	2.1	tracer injection	0.3	1,572 mg/L 2-propanol (K=0) 1,247 mg/L 1-pentanol (K=3.9) 1,144 mg/L 2-ethyl-1-butanol (K=12.5)
3.7	5.8	water flooding	2.6	
1.0	6.8	NaCl preflood	0.7	7,000 mg/L NaCl
3.4	10.2	surfactant/ alcohol flooding	2.4	7.55% sodium dihexyl sulfosuccinate 4.47 % isopropanol 7,000 mg/L NaCl
11.0	21.2	water flooding	7.8	
1.0	22.2	tracer injection	0.7	854 mg/L 1-propanol (K=0) 431 mg/L bromide (K=0) 798 mg/L 1-hexanol (K=30) 606 mg/L 1-heptanol (K=141)
5.1	27.3	water flooding	3.6	
2.4	29.7	extraction only	--	

Notes: All injected solutions are mixed in Hill tap water. The total injection rate was 1.7 m³/s (7.5 gpm) for all three injection wells and the total extraction rate was 2.1 m³/s (9.2 gpm) for all three extraction wells. The water injection rate for the hydraulic control well SB-8 was 1.6 m³/s (7 gpm).

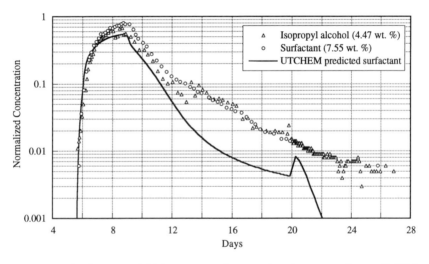

Figure 3. Surfactant and IPA concentrations produced at extraction well SB-1 during Phase II test: comparison of UTCHEM prediction with field data.

agreement with the field data, but we prefer to show the predictions that we made before the field test and comment that they were more than adequate to meet all of our stated objectives and design purposes. No comparable model predictions can be found in the literature. Thus, it is very worthwhile to show that even with all of the approximations and uncertainties inherent in such modeling, the predictions can be sufficiently accurate to be very useful for a variety of important purposes e.g. in estimating how much surfactant is needed to reduce the DNAPL saturation to a given level.

Over 94% of the surfactant was recovered and the final surfactant concentration in the effluent water was less than 0.05%. Figure 3 shows a normalized plot of the surfactant and isopropanol concentrations in the effluent. Breakthrough of surfactant and IPA occurred at the same time and no measurable separation occurred at any time, which indicates that there was negligible adsorption of the surfactant on the soil, a result consistent with the column studies.

Figure 4 shows the contaminant concentrations measured by GC analysis of extraction well fluid samples for the central extraction well SB-1 and compares these with UTCHEM predictions. The measured concentrations during the initial tracer test (first 5 days of the plot) are near the groundwater solubility of 1100 mg/L both in the prediction case and the field measured data. Shortly after surfactant injection begins at 5.5 days, the contaminant concentration increases steeply: to over 10,000 mg/L in the predictions and to over 20,000 mg/L in the field data. The decline in concentrations after surfactant injection ends at 8.7 days (2.4 PV) is slower in the field data than that for the UTCHEM prediction. This difference is at least partially due to the higher extraction rates in the predictive simulations, compared to those actually achieved in the field.

In general, field test results exhibit 'spiky' or non-smooth behavior in produced concentrations due to small scale fluctuations in rates and flow fields, sampling variations, measurement errors, heterogeneities, etc. The simulations similarly show 'spiky' behavior due to aquifer heterogeneities, spatially variable remaining DNAPL saturations, changes in the flow field, phase behavior changes, and the effect of structure upon each streamlines' arrival time at the effluent well. The 'spike' seen in the simulation at Day 20 is a result in a changing flow field, when all injection is stopped in the field and only extraction wells continue pumping, to maximize the recovery of any injected chemicals.

Figure 5 compares the difference between the water table depth (or fluid head) between each injection well and extraction well pair during the Phase II test. There was no loss of hydraulic conductivity during the Phase II surfactant flood, based on measured hydraulic gradients before and after surfactant injection. The fluid head levels increased slightly during the surfactant injection period due to the increased viscosity of the surfactant solution compared to water, but quickly returned to the pre-surfactant injection levels during the water flood. These and other laboratory and field data demonstrate that the soldium dihexyl sulfosuccinate surfactant is an extremely good choice for these conditions when used with a co-solvent such as isopropanol, which promotes microemulsions with very fast equilibration times, equilibrium solubilization and minimal surfactant adsorption on the soil.

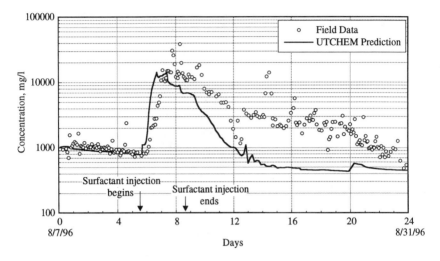

Figure 4. Contaminant concentrations produced at extraction well SB-1 during Phase II test: comparison of UTCHEM prediction with field data.

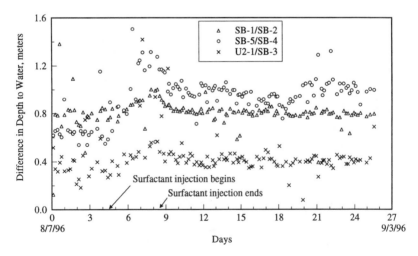

Figure 5. Difference in water table depth between test area extraction/injection well pairs.

Figure 6 shows the produced tracer concentrations for the final Phase II tracer test, conducted after the surfactant remediation. Even though very high partition coefficient tracers (K=30 and 141) were used for this test, very little retardation of these tracers are seen; in other words, effluent data for the different tracers overlay for the two outside extraction wells SB-5 and U2-1 and the central monitoring well SB-6 and show only a very slight difference in the tail for the central extraction well SB-1. Compare this to the substantial retardation observed for the 1-heptanol tracer (K=141) during the Phase I test in extraction well SB-1, shown in Figure 2 on the semi-log scale.

The DNAPL volumes and saturations determined from the first temporal moment of the effluent tracer concentrations (21,22) of the Phase II tracer test for the three extraction wells is summarized in Table III. The initial volume of DNAPL within the test pattern was 1310 L (6000 mg/kg of soil). The initial DNAPL volumes and saturations are based on moment analyses of the Phase I tracer test conducted before any surfactant injection. The final DNAPL volumes and saturations are based on moment analyses of the Phase II final tracer test conducted after the surfactant remediation. Values for each of the three extraction/injection well pairs and for the total test pattern are given in Table III. The initial DNAPL saturation ranged from 0.013 to 0.054, with an average of 0.027, equivalent to 6,000 mg/kg of soil averaged over the entire saturated volume of the aquifer.

After surfactant remediation, the average DNAPL saturation was 0.0003, a decrease of 99%, and the DNAPL saturations in the two outside swept volumes are too low to detect (less than 0.0001). This final average DNAPL saturation is equivalent to approximately 67 mg/kg of soil. After surfactant remediation, the amount of DNAPL remaining within the swept volume was only about 19 L. The estimated volume of DNAPL recovered based upon the effluent GC data taken on-site during the test was 1870 L. The estimated volume of DNAPL collected after steam stripping in the treatment plant was 1374 L.

We consider both of these DNAPL recovery estimates from the effluent data to be less accurate than the estimate of 1291 L from the tracer data. The partitioning tracer data do not depend on aquifer characteristics such as porosity and permeability. The estimated error is on the order of 12% of the estimated volume of DNAPL in the swept volume of the aquifer even when the DNAPL volume is very small provided the retardation factor is still sufficiently high. The partition coefficient must be high when the average NAPL saturation is low for the retardation factor to be sufficiently high. The final DNAPL volume estimate of 19 L reported above is based upon the 1-heptanol tracer data. The retardation factor for each well pair is uncertain by about ±0.035. This results in an uncertainty of 5 L of DNAPL for each well pair and a total of 15 L for the entire swept volume, or about 99±1% removal of the DNAPL from the swept volume of the aquifer.

The important conclusion is that these results clearly demonstrate that surfactant flooding can be used to remove essentially all of the DNAPL in the contacted volume of an aquifer, which is the source of the continuing contamination of the water for extended periods of time i.e. the large and mobile dissolved plume. Only three weeks were required to achieve this result. The total contaminant concentration in the central monitoring well at the end of the test was only 8 mg/L. This is a 99%

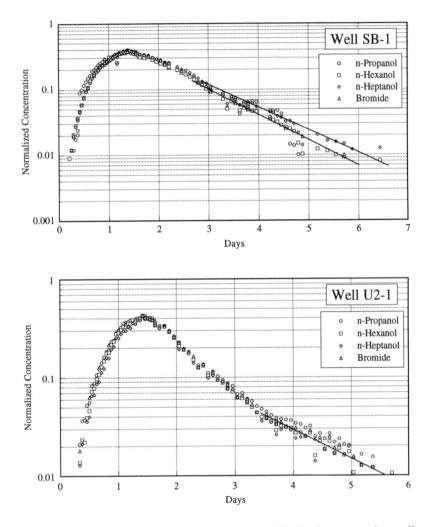

Figure 6a. Measured tracer concentrations produced in the three extraction wells and monitoring well during Phase II test, Wells SB-1 and U2-1.

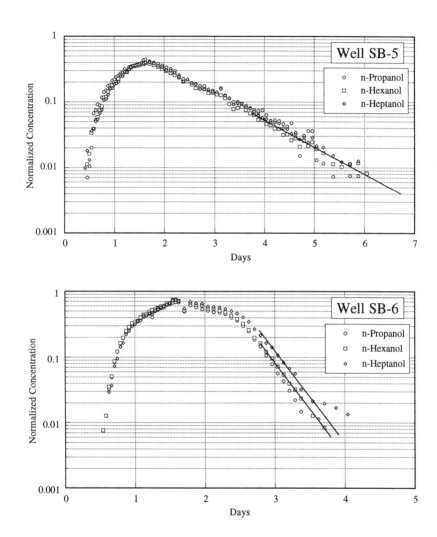

Figure 6b. Measured tracer concentrations produced in the three extraction wells and monitoring well during Phase II test, Wells SB-5 and SB-6.

Table III: Initial and Final DNAPL Volumes and Saturations from Tracer Tests

	Well Pair U2-1/SB-3	Well Pair SB-1/SB-2	Well Pair SB-5/SB-4	Total Swept within Test Area
DNAPL volume, liters				
Initial	250	795	265	1310
Final	0	19	0	19
DNAPL Saturation, %				
Initial	1.7	5.4	1.3	2.7
Final	0.00	0.10	0.00	0.03
Contaminant Soil Content				
Initial, mg/kg of soil	3,800	12,000	2,900	6,000
Final, mg/kg of soil	0	224	0	67

reduction compared to the initial concentrations. While this concentration is still much higher than the 0.005 mg/L MCL for TCE, a UTCHEM simulation showed that only 55 days of continued water flooding at the same well rates would be sufficient to reduce the aqueous concentration of the contaminants to the MCL (24). Either natural attenuation or some other means such as bioremediation could be used to degrade the contaminants remaining in the water now that almost all of the DNAPL source has been removed. However, the test area is an open geosystem and this final step could not be completed without either remediating the entire OU2 aquifer or isolating the remediated volume to prevent recontamination from outside the demonstration area.

Acknowledgments

The results described in this paper are partial results from a joint project between Duke Engineering and Services (formerly INTERA), Radian International LLC and The University of Texas. Many people at these organizations have contributed to the laboratory research, characterization of the site and the design and operation of the field tests including especially Sonny Casaus, Paul Cravens, Jeff Edgar, Stacy Griffith, Kurt Himle, Tim Hunsucker, Laureen Kennedy, Dino Kostarelos, Paul Mariner, Kermit Riley, Ken Rotert, Bruce Rouse, Ronald Santini, Douglas Shotts and Vinitha Weerasooriya. This project was sponsored by the Air Force Center for Environmental Excellence (AFCEE), Brooks AFB Texas and we wish to thank Sam Taffinder of AFCEE for his support of the project and Jon Ginn and other Hill Air Force base personnel for their great support and cooperation and the opportunity to conduct this test at Hill AFB. Financial support for modeling and laboratory experiments has been provided by both EPA and two State of Texas Advanced Technology Program grants that were crucial to the success of the project. We especially wish to thank Larry Britton of Condea VISTA chemicals for doing surfactant biodegradation studies for us during the research leading up to this field demonstration.

Literature Cited

(1) Mackay, D.M.; Cherry, J.A. *Environ. Sci. Technol.*, **1989**, *23*, 630-636.

(2) Soerens, T.S.; Sabatini, D.A.; Harwell, J.H. In Proceedings of the Subsurface Restoration Conference, Third International Conference on Ground Water Quality Research; EPA - R.S. Kerr Environmental Research Lab; Dallas, Texas, 1992, 269-270.

(3) Pennell, K.D.; Jin, M.; Abriola, L.M.; Pope, G.A. *Journal of Cont. Hydrology*, **1994**, *16*, 35-53.

(4) Dwarakanath, V., B. A. Rouse, G. A. Pope, D. Kostarelos, D. Shotts, and K. Sepehrnoori. "Surfactant Remediation of Soil Columns Contaminated by Nonaqueous Phase Liquids." accepted by *Journal of Contaminant Hydrology*, 1998.

(5) Hirasaki, G.J.; Miller, C.A.; Szafranski, R.; Tanzil, D.; Lawson, J.B.; Meinardus, H.; Jin, M.; Jackson, R.E.; Pope, G.; Wade, W.H. Presented at the 1997 SPE Annual Technical Conference and Exhibition; San Antonio, Texas, October 5-8, 1997.

84

(6) Knox, R. C.; Sabatini, D.A.; Thal, A. In Proceedings of the I&EC Special Symposium; Tedder, D.W., Ed.; American Chemical Society; Birmingham, Alabama, 1996, 116.

(7) Fountain, J.C.; Starr, R.C.; Middleton, T.; Beikirch, M.; Taylor, C.; Hodge, D. Ground Water, **1996**, *34*, 910-916.

(8) Abriola, L. M.; Dekker, T. J.; Pennell, K.D. *Environ. Sci. Technol.*, **1993**, *27*, 2341-2351.

(9) Brown, C. L.; Pope, G.A.; Abriola, L. M.; Sepehrnoori, K. *Water Resour. Res.,* **1994**, *30*, 2959-2977.

(10) Pope, G. A.; Jin, M.; Dwarakanath, V.; Rouse, B. A.; Sepehrnoori, K. "Partitioning Tracer Tests to Characterization Organic Contaminants," in Proceedings of the Second Tracer Workshop, Bjørnstad, T.; Pope, G. A., Eds.; Austin, Texas, 1994.

(11) Annable, M.D.; Jawitz, J.W.; Rao, P.S.C.; Dai, D.P.; Kim, H.; Wood, A.L. *Ground Water*. **1998**, *36*, 495-502.

(12) Nelson, N.T.; Brusseau, M.L.*Environ. Sci. Technol.*, **1996**, *30*, 2859-2863.

(13) Studer, J.E.; Mariner, P.; Jin, M.; Pope, G.A.; McKinney, D.C.; Fate, R. "Application of a NAPL Partitioning Interwell Tracer Test (PITT) to Support DNAPL Remediation at the Sandia National Laboratories/New Mexico Chemical Waste Landfill," in Proceedings of the Superfund/Hazwaste West Conference, Las Vegas, Nevada, 1996.

(14) Oolman, T.; Godard, S. T.; Pope, G. A.; Jin, M.; Kirchner, K. *GMWR*, 1995, 125-137.

(15) Radian. Final Remedial Investigation for Operable Unit 2-Sites WP07 and SS21, Prepared for Hill AFB, Utah. 1992.

(16) Intera. Phase I Work Plan, Hill Air Force Base, 00-ALC/EMR. 1996.

(17) Radian. Aquifer Data Evaluation Report, Operable Unit 2, Hill Air Force base, Utah. Radian, RCN 279-100-26-41, 1994.

(18) Baran, Jr., J.R.; Pope, G.A.; Wade, W.H.; Weerasooriya, V.; Yapa, A. . *Environ. Sci. Technol.*, **1994**, *28*, 1361-1366.

(19) Dwarakanath, V. "Characterization and Remediation of Aquifers Contaminated by Nonaqueous Phase Liquids Using Partitioning Tracers and Surfactants," May, 1997, Ph.D. Dissertation, U. of Texas, Austin.

(20) Jin, M.; Delshad, M.; Dwarakanath, V.; McKinney, D.C.; Pope, G.A.; Sepehrnoori, K.; Tilburg, C.; Jackson, R.E. *Water Resour. Res.*, **1995**, *30*, 1201-1211.

(21) Jin, M. "A Study of Nonaqueous Phase Liquid Characterization and Surfactant Remediation," 1995, Ph.D. Dissertation, U. of Texas, Austin.

(22) Pope, G. A.; Sepehrnoori, K.; Delshad, M.; Rouse, B. A.; Dwarakanath, V.; Jin, M. "NAPL Partitioning Interwell Tracer Test in OU1 Test Cell at Hill Air Force Base, Utah," Final Report, ManTech Environmental Research Services Corp., P.O. number 94RC0251, GL number 2000-602-4600, 1994.

(23) Results of the Surfactant Treatability Study for Operable Unit 2 Hill Air Force Base, Utah, Radian RCN 279-104-08-10, 1994.

(24) Demonstration of Surfactant Enhanced Aquifer Remediation of Chlorinated Solvent DNAPL at Operable Unit 2, Hill AFB, Utah, Final Report prepared by

INTERA for the Air Force Center for Environmental Excellence, Technology Transfer Division, January, 1998.

(25) Delshad, M.; Pope, G.A.; Sepehrnoori, K. *Journal of Cont. Hydrology,* **1996**, *23*, 303-327.

(26) Brown, C.L. "Design of a Field Tests for Surfactant-Enhanced Remediation of Aquifers Contaminated by Dense Nonaqueous Phase Liquids," Ph.D. Dissertation, U. of Texas, Austin, 1998.

Chapter 7

In-Situ Solubilization by Cosolvent and Surfactant-Cosolvent Mixtures

Michael D. Annable[1], James W. Jawitz[1], Randall K. Sillan[2], and P. Suresh Rao[2]

[1]Departments of Engineering Sciences and [2]Soil and Water Science, University of Florida, Gainesville, FL 32611

Introduction

This chapter summarizes the results of two studies conducted by the University of Florida at Operable Unit 1 (OU1) at Hill AFB, Utah. Both studies involved injection of an agent designed to solubilize rather than mobilize a multicomponent non-aqueous phase liquid (NAPL) that was lighter than water. In these studies, mobilization was avoided because of uncertainty associated with control of the mobilized phase and the potential transport of the NAPL into previously uncontaminated regions of the aquifer. While solubilization is less efficient, the increased efficiency gained during mobilization was deemed not worth the risk of increased contamination. The first study, conducted in 1995, employed an alcohol mixture that was flushed through the NAPL-contaminated aquifer. The second study, conducted in 1996, employed a surfactant/alcohol mixture designed to microemulsify the NAPL. In this chapter, brief summaries and comparisons are made between the two studies. For more details on each study, the reader is referred to Rao et al. (*1*) for the alcohol flushing study, and Jawitz et al. (*2*) for the surfactant/alcohol flushing study. In this chapter, the first study will be referred to as the Cosolvent study and the second will be referred to as the Single-Phase Microemulsion (SPME) study.

Both studies were conducted as part of a larger project designed to compare nine technologies side-by-side. This effort was coordinated by Carl Enfield and Lynn Wood at the National Risk Management Laboratory in Ada, Oklahoma. Primary funding for the comparison study came from the Strategic Environmental Research and Development Program (SERDP). All the technology studies followed similar protocols to provide a common basis for evaluation. The next section describes the two technologies used in the studies conducted by the University of Florida, providing references to relevant literature for further details. This is followed by sections presenting summaries of the methods common to the two technologies and a comparison of the data collected during the two studies.

86

Technology Summaries

Cosolvent Flushing. The *in-situ* cosolvent flushing technique provides accelerated site cleanup due to enhanced solubilization or mobilization of NAPL, and enhanced desorption of sorbed contaminants (*1,3-12*). The concept of using cosolvents for remediation of contaminated soils first appeared in the literature in the mid-80's in studies investigating organic contaminant sorption processes (*4,13-17*). Prior to this work, theory on cosolvency relationships had been developed in pharmaceutical research (e.g., Yalkowsky and Rosen (*18*)). The method had originally been advanced in the petroleum research community as a tool for enhanced oil recovery from reservoirs (*19-21*). Over the last decade, the concept has been expanded with more recent studies focusing on enhanced removal of sorbed organic contaminants (*8-10,22*), and dissolution or mobilization of residual NAPLs (*5-7,12,23-26*). The first field test of cosolvents for solubilization of NAPL was conducted at a controlled release site (*28*). The first comprehensive evaluation of cosolvent remediation at a contaminated field site was recently reported by Rao et al. (*1*).

The cosolvent flushing process provides solubility enhancement because addition of organic solvents to water decreases the polarity of the mixture. The decreased polarity increases the solubility of nonpolar organic compounds present in NAPLs. The solubility of a compound in a cosolvent mixture (S_m), follows a log-linear relationship (*29*):

$$\log S_m = \log S_w + \sum \beta_i \sigma_i f_{c,i}$$

where S_w is the solubility of the compound in water, f_c is the volume of cosolvent i, σ_i is the cosolvency power, and β_i is an empirical coefficient that accounts for water-cosolvent interactions. For multi-component NAPLs, Raoult's law can be used to predict the concentration in solution (C_i):

$$C_i = \chi_i S_i$$

where χ_i is the mole fraction of the compound in the NAPL, and S_i is the liquid solubility of the compound in water. Since the liquid solubility increases in a log-linear fashion with cosolvent content, it follows that C_i obeys the same log-linear relationship (*9,10*).

Based on the cosolvency theory briefly outlined above and laboratory experiments conducted using various alcohol mixtures (*30*), a ternary mixture of water and two alcohols was selected for the cosolvent flood at the Hill AFB test site. Ethanol was the primary flushing cosolvent (70% volume fraction), and *n*-pentanol (12%) was added to enhance solubilization of the strongly hydrophobic constituents present in the NAPL.

In the field test, an ethanol-water binary mixture was injected into the test cell using a gradient injection program linearly increasing the alcohol content over approximately one-half a pore volume. This was done because flow instabilities, caused by differences in fluid properties, were a concern during both cosolvent flushing and subsequent water flushing (*7,11,22,31*). The gradient injection was followed by a step-change addition of a ternary mixture of ethanol, pentanol, and water. The ternary alcohol mixture was applied over a 10-day period corresponding to approximately nine test-cell pore volumes. The ternary flood was followed by a one pore volume ethanol/water flood that was then transitioned to water using a linear gradient injection program. During cosolvent flushing, and subsequent water flushing, samples were collected at three extraction wells and 60 multi-level samplers.

SPME Study. Like the cosolvent solubilization study, the SPME approach was designed to remove NAPL mass without mobilization as defined earlier. The method uses an injected solution designed to produce a microemulsion upon contact with NAPLs. A microemulsion is defined as an optically transparent dispersion of liquid droplets (< 0.1 μm) that is suspended within a second, immiscible liquid and is stabilized by an interfacial film of surface-active molecules (*32*). A cosurfactant, typically intermediate chain-length alcohols, is often added to the surfactant. Adjusting physical or chemical properties can change the character of a system from hydrophilic to lipophilic producing a phase transition from an oil-in-water microemulsion (oil, or NAPL, droplets in a water-continuous phase; Winsor Type I), to a water-in-oil microemulsion (water droplets in an oil phase; Winsor Type II). A Winsor Type III system may form mid-way in the transition from a hydrophilic to lipophilic system. At this point, the interfacial tension between the microemulsion phase (or middle phase), and any excess oil or water phases present reach a minimum value. It is because of these ultra-low interfacial tensions that petroleum engineers have embraced middle-phase microemulsions for enhanced oil recovery processes (*33,34*).

The SPME approach employed at Hill AFB used a Winsor Type I microemulsion to affect mass removal of NAPL. Martel and collaborators (*35,36*) first proposed the use of Winsor Type I microemulsions to solubilize NAPLs without mobilization. These systems have the advantages of high solubilization of NAPLs (although lower than middle-phase microemulsions) with relatively low amounts of chemical additives required. Chun-Huh (*37*) demonstrated that for microemulsions, solubilization of the oil phase is related to interfacial tension by an inverse-squared relationship. Based on this relationship, Winsor Type I microemulsification will necessarily be less efficient for remediation than Winsor Type III microemulsions. In order to avoid mobilization, interfacial tension must be higher and therefore solubilization will be reduced.

During the surfactant-cosurfactant selection phase of the SPME study at Hill AFB, eighty-six surfactants and a number of alcohols were screened, with maximum NAPL solubilization and low-viscosity (< 2 cp) as the main acceptance criteria (*38*). The viscosity of the precursor solution was limited to preclude large hydraulic gradients across the test cell and excessive fluid level rise in the injection wells and drawdown around the extraction wells. The precursor solution used in the field study was the surfactant Brij 97® (polyoxyethylene (10) oleyl ether) at 3% by weight and *n*-pentanol at 2.5% by weight in water. Rhue et al. found that equilibration with the Hill AFB NAPL produced dissolved concentrations of dodecane (approximately 0.7% by weight of the NAPL) of 104 mg/L, compared to an aqueous solubility of 0.0037 mg/L.

The field-measured viscosity of the precursor was 1.66 cp. The precursor solution was also tested prior to the field experiment in flushing experiments conducted at the laboratory scale using contaminated soil from the field site in one-dimensional flow columns (*38*). While greater NAPL solubilization could be achieved with higher concentrations of surfactant and cosurfactant, increased solubilization would be accompanied by a potentially unacceptable increase in viscosity.

The field implementation phase of the SPME flushing test was similar to the cosolvent test described above. The SPME flood employed approximately nine pore volumes of the precursor. Gradient injections were not used because the viscosity and density of the precursor was selected such that bulk fluid properties would be relatively compatible with the resident groundwater. During the precursor displacement by water, a one pore volume pulse of a solution containing only the surfactant was used to more effectively remove the *n*-pentanol.

General Experimental Approach

The goal of the field studies was to remove NAPL from the contaminated aquifer and quantify this using multiple evaluation methods (*39*). The studies were conducted in isolation test cells to provide hydraulic control and allow for more accurate mass balance calculations (Figure 1). Each test was evaluated using: 1) Pre- and post-flushing soil core sample analysis, 2) Pre- and post-flushing partitioning tracer tests, 3) Effluent mass removal compared to initial mass estimates, 4) Pre- and post-flushing groundwater sample analysis.

To collect the data required for these comparisons, a sequence of experiments and sampling events were performed. Following installation of the Waterloo sealable sheetpile test cells (*40*), soil core samples were collected during installation of the injection/extraction wells and multi-level samplers. A total of 105 and 68 samples were collected from the cosolvent and SPME studies, respectively. In the field, approximately 10 g of soil were transferred into vials containing 5 ml of methylene-chloride for extraction. These samples were analyzed by GC-MS for a suite of target analytes present in the NAPL (for more details see (*1,2*)). These samples were used to estimate the total mass of each analyte in the test cell.

Following installation of the wells and multi-level samplers and before establishing flow in the test cell, ground water samples were collected and analyzed for the target analytes. Following this, a preliminary non-reactive tracer test was conducted to characterize the hydrodynamics, determine the swept volume, and estimate the porosity (*40*). Breakthrough curves (BTCs) from this experiment were also used to set sampling schedules for future tests. Bromide and iodide were used for the non-reactive tracer tests for the cosolvent and SPME studies, respectively.

The non-reactive tracer test was followed by a partitioning tracer test (*41-45*). The goal of this test was to quantify the volume of NAPL present in the swept zone and, using the multi-level samplers, characterize the spatial distribution. Both studies, cosolvent and SPME, used a suite of partitioning and non-partitioning tracers but the analysis focused primarily on bromide and methanol (non-reactive) and 2,2-dimethyl-3-pentanol (partitioning coefficient of approximately 12).

Figure 1.   Cosolvent test site.

The flushing phase of each study was conducted after the results of the pre-flushing partitioning tracer test were evaluated. The tests involved flushing with approximately 9 pore volumes of flushing agent. The average flow rate was approximately one pore volume per day (approximately 4 m/day). Each study included flow perturbations, either flow-interruption periods, or periods with reduced flow rates to evaluate non-equilibrium dissolution. During each flushing study, approximately 5000 samples were collected from the extraction wells and multi-level samplers. After the 9 pore volumes, the flushing agent was displaced with water. About 20 and 10 pore volumes of water were displaced prior to the post-flushing partitioning tracer test for the cosolvent and SPME studies, respectively.

The final test of each study was the post-flushing partitioning tracer test. This was conducted under the same flow conditions using the same tracers as the pre-flushing test. The goal was to determine the amount and distribution of NAPL for comparison with the pre-flushing results. Upon completion of the post-flushing partitioning tracer test, and following a 72-hour equilibration period, ground water samples were collected for comparison with the pre-flushing samples. Finally, 60 and 36 soil core samples were collected from within the swept zones for the cosolvent and SPME studies, respectively.

All the samples collected were analyzed and the data compiled to generate BTCs from which temporal moments were calculated. Zeroth moments were used to determine mass removal (or recovery of tracers). First moments allowed determination of mean arrival times, which were used to calculate retardation of partitioning tracers and swept volumes based on non-reactive tracers. Concentrations of target analytes in soil samples were averaged to compare pre-and post-flushing values, and to estimate total contaminant (using an assumed value for the average soil bulk density).

Comparative Results

The following sections summarize the results of the cosolvent and SPME studies and present limited data sets from each study for comparison. The results presented are followed by a discussion comparing the performance of each technology tested.

Soil Core Samples. Soil core samples, collected prior to each flushing test, were analyzed for a suite of compounds present in the NAPL. Following the final post-flushing partitioning tracer test, soil samples were again collected and analyzed for the same constituents. Pre- and post-flushing concentrations of n-decane in soils samples are plotted over the vertical dimension of the test cell for both flushing tests (Figure 2). Both experiments showed substantial removal of n-decane as a result of the in situ flushing process. Average concentrations of n-decane, based on samples collected in the flushing swept zone, are provided in Table 1 and indicate that the cosolvent and SPME studies produced fractional removals of 0.89 and 0.96, respectively.

Partitioning Tracers. Partitioning tracers tests were conducted before and after the in situ flushing tests. Pre- and post-flushing BTCs of non-partitioning and partitioning tracers from the central extraction well of each test cell are presented in Figure 3. The

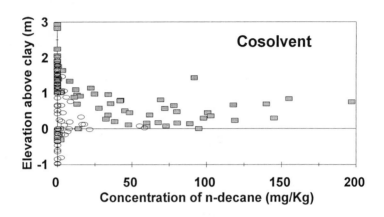

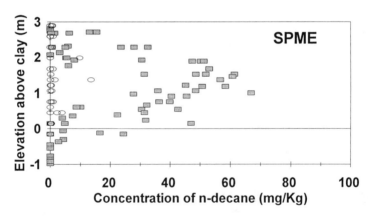

Figure 2. Pre- and post-flushing soil core *n*-decane concentrations. (Pre-flushing ■, Post-flushing ○).

Table 1. Pre- and post-flushing mass removal estimates for both the cosolvent and SPME in situ flushing tests.

Method	Cosolvent			SPME		
	Pre-flushing	Post-flushing	Fraction Reduction	Pre-flushing	Post-flushing	Fraction Reduction
Soil Cores[1] (n-decane)	130 g/m²	14 g/m²	0.89	24.9 mg/Kg	1.10 mg/Kg	0.96
NAPL saturation at EW2 based on partition-ing tracer (DMP)	0.049	0.009	0.81	0.068	0.015	0.78
n-decane based on Extraction Wells	(estimate[2]) 2.07 Kg	(removed) 1.86 Kg	0.90	(estimate[2]) 1.87 Kg	1.12 Kg	0.61
DCB in Ground Water	1100 µg/L	127 µg/L	0.88	--	--	--

1 – Soil core averages were calculated on a mass per square surface area basis for the cosolvent test cell while the SPME cell was an arithmetic average of all samples.

2 – pre-flushing mass estimates based on concentrations measured in soil cores

DMP – tracer 2,2-dimethyl-3-pentanol

DCB - of 1,2-dichlorbenzene

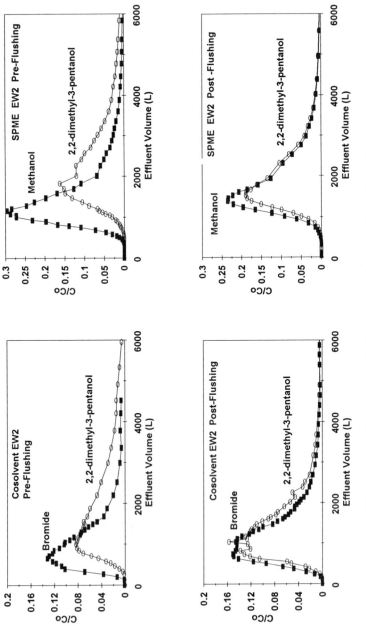

Figure 3.   Pre- and post-flushing partitioning tracer breakthrough curves.

central extraction well was selected because it is expected to be influenced the least by the corrugations of the test-cell sheet-pile walls. The reduced retardation of the partitioning tracer 2,2-dimethyl-3-pentanol (DMP) following in situ flushing indicates substantial NAPL removal in each test. The pre- and post-flushing NAPL saturations based on the central extraction well (EW2) are reported in Table 1. The NAPL removal effectiveness of both studies was comparable (0.81 and 0.78 for the cosolvent and SPME tests, respectively).

Extraction Well Mass Balance. Extraction well samples were collected during the in situ floods and subsequent water floods to quantify the NAPL constituent mass removal. A photograph of extraction well samples collected during the SPME flood indicates a substantial color change between the first and second pore volumes following the initiation of the precursor introduction (Figure 4). The color change observed qualitatively agrees with concentration of *n*-decane measured in the samples (Figure 5). The breakthrough behavior and *n*-decane concentrations observed in both studies demonstrate the substantial solubility enhancement that the tests are designed to produce (aqueous solubility of *n*-decane is approximately 0.003 mg/L.

The mass removal from each test cell was determined by calculating zeroth moments of the analyte BTCs (Table 1). The mass of n-decane removed was compared to the estimated mass of n-decane initially present in the test cell. The estimate of initial mass presented in Table 1 is based on the pre-flushing soil core analysis and an assumed bulk density of the media ($\rho_B = 1.7$ g/cm^3). The resulting recoveries, 0.90 and 0.61 for the cosolvent and SPME studies, respectively, suggest that there may be high uncertainty in the initial mass estimate. When comparing two mass calculations that based on very different measurement techniques, such as core analysis and effluent sample analysis, uncertainties and biases in each measure could lead to significant errors. Whereas, measurements determined from the same technique, such as pre- and post-core analysis and pre- or post-partitioning tracers, provide consistent measures that should lead to lower errors in mass removal calculations.

Ground Water Samples. Ground water samples were collected before and after each flushing study. These samples were analyzed for the study target analytes to characterize changes in ground water quality. Samples were collected prior to initiating flow in the test cell to characterize the existing conditions. The post-flushing samples were collected at least 72 hours following the final partitioning tracer test. The average concentration of 1,2-dichlorbenzene in the cosolvent test cell is presented in Table 1 (*n*-decane was not used because of the low solubility in water). In the cosolvent flushing study the reduction in the average concentration of DCB within the test cell was about 90%. This level of concentration reduction is comparable of other constituents monitored and is in agreement with other measures of performance (*1*). In the SPME test cell, the initial concentrations of DCB in ground water samples were nearly all below the detection limit (10 μg/L). This was observed for all constituents monitored. As expected post-flushing samples were below detection as well. The reason for the difference between the two test cells is likely due to the location of each cell relative to the NAPL plume and general ground water flow direction. The SPME

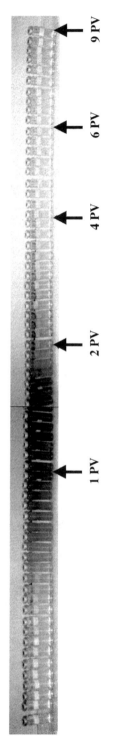

Figure 4.   Effluent sample vials from the SPME-flushing test.

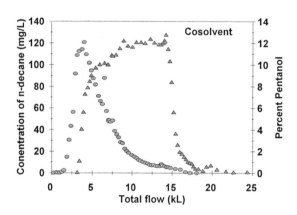

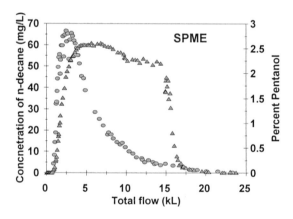

Figure 5.   Concentration of *n*-decane during the in-situ flushing test (n-decane •, Pentanol ▲).

cell was located near the leading edge of the NAPL plume where years of dissolution by groundwater had depleted the more water-soluble components in the multicomponent NAPL. In contrast, the cosolvent test cell was located near the down-gradient edge of the NAPL plume, leading to local enrichment of the NAPL in the more soluble components. This observation was supported by constituent analysis of NAPL collected from each test cell (1,2). DCB had a mass fraction of 0.0061 in the cosolvent test cell and approximately 0.0002 in the SPME cell. The issue of differences in NAPL composition and implications for comparing technologies is discussed in the following section.

Discussion

The comparisons made between the cosolvent and SPME floods in the previous section are useful as gross comparisons and indicate that the NAPL removal effectiveness of the two technologies was comparable. The differences observed between the two tests are likely not significant enough to conclude that one technology is superior to the other. Comparison of any two technologies must be viewed carefully considering differences in test cells that could contribute to the variations observed in the results. In an ideal study, the two test cells would have identical initial conditions. The overall SERPD project plan to conduct multiple studies within the same NAPL plume was aimed at approximating identical initial conditions. This is, of course, impossible in any real geologic media with a NAPL plume of limited extent. The OU1 site at Hill AFB may be quite good in this respect having a NAPL plume of 3.5 ha with a relatively limited NAPL smear zone. At OU1, however, the confining unit into which the sheet pile walls forming each test cell were driven was not flat, giving rise to significant differences in saturated zone thickness within each test cell and influencing the relative position of the NAPL within this zone. For example, in the SPME test cell the saturated thickness of the flood was twice that for the cosolvent test cell. This, along with the fact that the degree of NAPL heterogeneity could vary significantly between test cells, results in each cell having different initial conditions. Evidence of NAPL compositional variability, as presented earlier, also provides different initial conditions. These differences can lead to several processes that impact the validity of comparing technologies.

The relative position of the NAPL within the flushing-technology flow field could produce conditions more favorable to one technology. An example might be a system with NAPL primarily located in the upper portion of the flow domain which could favor a cosolvent flood which uses a flushing agent with a density less than water. On the other hand, a system with NAPL located predominantly near the confining unit, such as the case for the cosolvent flood at Hill, could prove detrimental for a cosolvent flood. Even considering heterogeneities of the NAPL alone, irrespective of the flushing technology characteristics, systems where the NAPL is heterogeneously distributed in space, such as a DNAPL pool, will be much more difficult to extract than a more uniformly distributed NAPL. The point is made here only to suggest that further investigation of the spatial pattern of the NAPL distribution and the spatial pattern of technology performance should be considered when making technology comparisons.

Differences in NAPL composition can also impact the ultimate effectiveness of a technology. A NAPL that is more hydrophobic, possibly as a result of significant dissolution by groundwater or length of the aging process, will likely be more challenging to extract, particularly for technologies which rely on solubilization. One can make appropriate corrections for differences in mass fractions for a single component in a multicomponent NAPL, as suggested by Jawitz et al. (*46*). However, approaches need to be developed that account for variability in the hydrophobicity of the remaining NAPL mixture. A difference in wettability of the NAPL is another characteristic that could lead to significant differences between studies. Systems that have become oil-wet may have very different NAPL-aqueous phase contact as compared to water-wet systems.

Summary

A comparison of two in-situ flushing studies conducted by the University of Florida at Hill AFB, Utah indicates that both remediation technologies were successful at removing a significant fraction of the initial contamination present in the surficial aquifer. The tests compared were both in-situ flushing tests, one employing a combination of two alcohols (cosolvent study) the second employing a combination of a surfactant and an alcohol (SPME study). Both flushing methods removed approximately 90% of the mass based on soil cores and constituent removal in extraction wells and approximately 80% based on partitioning tracers. Differences between the results presented here were not substantial and do not warrant identifying one technology over the other. Further spatial analysis of the two experiments may reveal important differences in the initial conditions that could lead to favoring one technology over another.

Literature Cited

1. Rao, P.S.C.; Annable, M.D.; Sillan, R.K.; Dai, D.P.; Hatfield, K.; Graham, W.D.; Wood, A.L.; Enfield, C.G. *Water Resour. Res.* **1997**, 33 (12), 2673-2686.
2. Jawitz, J.W.; Annable, M.D.; Rao, P.S.C.; Rhue, R.D. *Environ. Sci. Technol.* **1998**, 32 (4), 523-530.
3. U.S. EPA. 1991. *In-situ* solvent flushing. EPA/540/2-91/021. Office of Emergency Response, U.S. EPA, Washington, DC.
4. Rao, P.S.C.; Lee, L.S.; Wood, A.L. 1991. Solubility, Sorption, and Transport of Hydrophobic Organic Chemicals in Complex Mixtures. EPA/600/M-91/009, R.S. Kerr Env. Res. Lab., U.S. EPA, Ada, OK.
5. Luthy, R.G.; Dzombak, D.A.; Peters, C.A.; Ali, M.A.; Roy, S.B. 1992. Solvent extraction for remediation of manufactured gas plant sites. EPRI/TR-101845. Final Research Project (3072-02) Report, Electric Power Research Institute, Palo Alto, CA.
6. Rixey, W.G.; Johnson, P.C; Deely, G.M.; Byers, D.L.; Dortch, I.J. 1992. Mechanisms of removal of residual hydrocarbons from soils by water, solvent, and surfactant flushing. Chapter 28, In: *Hydrocarbon Contaminated Soils*, Lewis Publishers, Boca Raton, FL.

100

7. Farley, K.J.; Falta, R.W.; Brandes, D.; Milazzo, J.T.; Brame, S.E. 1993. Remediation of hydrocarbon contaminated groundwaters by alcohol flooding. Technical Report submitted to Hazardous Waste Management Research Fund, Univ. of South Carolina, SC.
8. Augustijn, D.C.M.; Jessup, R.E.; Rao, P.S.C.; Wood, A.L. *J. Environ. Eng.* **1994**, 120 (1), 42-57.
9. Augustijn, D.C.M.; Dai, D.; Rao, P.S.C.; Wood, A.L. 1994. Solvent flushing dynamics in contaminated soils. pp. 557-562, In: *Transport and Reactive Processes in Aquifers* (eds., Th. Dracos and F. Stauffer), A.A. Balkema, Rotterdam. The Netherlands.
10. Augustijn, D.C.M.; Lee, L.S.; Jessup, R.E.; Rao, P.S.C.; Annable, M.D.; Wood, A.L. 1997. Remediation of soils contaminated with hydrophobic organic chemicals: Theoretical basis for the use of cosolvents. Chapter 15, In: *Subsurface Restoration Handbook* (eds., C.H. Ward, J.A. Cherry, and M.R. Scalf), Ann Arbor Press, Inc., Chelsea, MI. (In Press).
11. Grubb, D.C.; Sitar, N. 1994. Evaluation of technologies for in-situ cleanup of DNAPL contaminated sites. EPA/600/R-94/120, R.S. Kerr Environmental Research Lab., U.S. EPA, Ada, OK.
12. Imhoff., P.T.; Gleyzer, S.N.; McBride, J.F.; Vancho, L.A.; Okuda, I.; Miller, C.T. *Environ. Sci. Technol.* **1995**, 29 (8), 1966-1976.
13. Rao, P. S. C.; Hornsby, A.G.; Kilcrease, D.P.; Nkedi-Kizza, P. *J. Environ. Qual.* **1985**, 14, 376-383.
14. Fu, J.K.; Luthy, R.G. *J. Environ. Eng.* **1986**, 12 (2), 328-345.
15. Fu, J.K.; Luthy, R.G. *J. Environ. Eng.* **1986**, 12 (2), 346-366.
16. Nkedi-Kizza, P.; Rao, P.S.C.; Hornsby, A.G. *Environ. Sci. Technol.* **1985**, 19 (10), 975-979.
17. Nkedi-Kizza, P.; Rao, P.S.C.; Hornsby, A.G. *Environ. Sci. Technol.* **1987**, 21 (11), 1107-1111.
18. Yalkowsky, S.H.; Roseman, T. 1981. Solubilization of drugs by cosolvents. In: *Techniques for Solubilization of Drugs*. S.H. Yalkowsky (ed.), Marcel Dekker, Inc., NY. pp. 91-134.
19. Gatlin, C.; Slobod, R.L. The alcohol slug process for increasing oil recovery. *Trans. of AIME*, 219:46-53. [20]   Taber, J.J., I.S.K. Kamath, and R.L. Reed. 1961. Mechanism of alcohol displacement of oil from porous media. *Trans. of AIME, 222: 195.*
21. Holm,L.W.; Csaszar, A.K. 1962. Oil recovery by solvents mutually soluble in oil and water. *Trans AIME, Vol 225, pp.129.*
22. Wood, A.L. 1995. Influence of cosolvents on the transport of hydrophobic organic chemicals in   soils under isocratic and gradient conditions. Ph.D. Dissertation, University of Oklahoma, Norman, OK.
23. Boyd, G.R.; Farley, K.J. 1992. NAPL removal from groundwater by alcohol flooding: Laboratory studies and applications. In: *Hydrocarbon Contaminated Soils and Groundwater*, Vol.. 2, E.J. Calabrese and P.T. Kostecki (eds.), Lewis Publishers, Boca Raton, FL.
24. Peters, C.A.; Luthy, R.G. *Environ. Sci. Technol.* **1993**, 27 (13), 2831-2843.
25. Peters, C.A.; Luthy, R.G. *Environ. Sci. Technol.* **1994**, 28(7):1331-1340.
26. Roy, S.B.; Dzombak, D.A.; Ali, M.A. *Water and Environ. Res.* **1995**, 67 (1), 4-15.
27. Hayden, N.J.; Vander Haven, E.J. *Water Env. Feder.*, **1996**, 68(7):1165-1171.
28. Broholm, K.; Cherry, J.A. 1994. Enhanced dissolution of heterogeneously distributed solvents residuals by methanol flushing: A field experiment. pp. 563-568, In: *Transport and Reactive Processes in Aquifers* (Th. Dracos & F. Stauffer, Eds), A.A. Balkema, Rotterdam, The Netherlands.
29. Morris, K.R.; Abramowitz, R.; Pinal, R.; Davis, P.; Yalkowsky, S.H. *Chemosphere*, **1988**, 17, 285-298.

30. Johnson, G.R. 1996. Solvent flushing of a LNAPL contaminated soil: Comparison of two cosolvent mixtures and the effects of weathered contamination on cosolvent flushing efficiency. M.E. Thesis, . Department of Environmental Engineering Sciences, University of Florida, 182pp.

31. Jawitz, J.W.; Annable, M.D.; Rao, P.S.C. *J. Contam. Hydrol.*, **1998**, 31 (3-4), 1-20.

32. Rosen, M.J. 1989. Surfactants and Interfacial Phenomena. John Wiley & Sons, New York, 431 pp.

33. Shah, D.O. (Editor), 1981. Surface Phenomena in Enhanced Oil Recovery. Plenum Press, New York, 874 pp.

34. Lake, L.W. 1989. *Enhanced Oil Recovery.* Prentice Hall. Englewood Cliffs, NJ.

35. Martel, R.; Gélinas, P.J. *Ground Water.* **1996**, 34 (1), 143-154.

36. Martel, R.; Gélinas, P.J.; Desnoyers, J.E.; Masson, A. *Ground Water.* **1993**, 31 (5), 789-800.

37. Huh, C. 1979. *J. Colloid Interface Sci.*, **1979**, 71, 408.

38. Rhue, R.D.; Annable, M.D.; Rao, P.S.C. 1997. Lab and field evaluation of single-phase microemulsions (SPME) for enhanced in-situ remediation of contaminated aquifers. Phase I: Laboratory studies for selection of SPME precursors, AATDF Report, University of Florida, Gainesville, FL.

39. Annable, M.D.; Rao, P.S.C.; Sillan, R.K.; Hatfield, K.; Graham, W.D.; Wood, A.L.; Enfield, C.G. Field-Scale application of *in-situ* cosolvent flushing: Evaluation approach, Proceedings of the Non-Aqueous Phase Liquid Conference ASCE, Washington D.C., Nov. 1996, pp. 212-220.

40. Starr, R.C.; Cherry, J.A.; Vales, E.S. 1992. A new steel sheet piling with sealed joints for groundwater pollution control, paper presented at the 45th Canadian Geotechnical Conference, Can. Geotech. Soc., Toronto, Ont., Oct. 26-28.

41. Annable, M.D.; Rao, P.S.C.; Graham,W.D.; Hatfield, K.; Wood, A.L. *J. Environ. Eng.* **1998**, 124 (6), 498-503.

42. Jin, M.; Delshad, M.; Dwarakanath, V.; McKinney, D.C.; Pope, G.A.; Sepehrnoori, K.; Tilburg, C.; Jackson, R.E. *Water Resour. Res.*, **1995**, 31 (5), 1201-1211.

43. Annable, M.D.; Rao, P.S.C.; Graham, W.D.; Hatfield, K.; Wood, A.L. Use of partitioning tracers for measuring residual NAPL distribution in a contaminated aquifer: Preliminary results from a field-scale test, in *2nd Tracer Workshop*, 77-85, Austin, TX, 1995.

44. Nelson, N. T. and M. L. Brusseau. 1996. Field study of the partitioning tracer method for detection of dense nonaqueous phase liquid in a trichloroethene-contaminated aquifer, *Environ. Sci. Technol.*, 30(9): 2859-2863.

45. Wilson, R. D. and D. M. Mackay. 1995. Direct detection of residual nonaqueous phase liquid in the saturated zone using SF6 as a partitioning tracer, *Environ. Sci. Technol.*, 29(5), 1255-1258.

46. Jawitz, J.W.; Sillan, R.K.; Annable, M.D.; Rao, P.S.C. 1997. Methods for determining NAPL source zone remediation efficiency of in-situ flushing technologies; In-Situ Remediation of the Geoenvironment: Proceedings of the Conference, Geotechnical Special Publication No. 17, ASCE: 271-283.

Chapter 8

Design and Performance of a Field Cosolvent Flooding Test for Nonaqueous Phase Liquid Mobilization

Ronald W. Falta[1,2], Cindy M. Lee[2], Scott E. Brame[1], Eberhard Roeder[2], Lynn Wood[3], and Carl Enfield[3]

[1]Department of Geological Sciences, Brackett Hall, Box 341908, Clemson University, Clemson, SC 29634–1908
[2]Department of Environmental Engineering and Science, Rich Laboratory at Clemson Research Park, Clemson University, Clemson, SC 29634–0919
[3]National Risk Management Research Laboratory/Subsurface Protection and Remediation Division at Ada, U.S. Environmental Protection Agency, 919 Kerr Research Drive, Ada, OK 74820

A pilot scale field test of the removal of a complex mixture of hydro-carbons, chlorinated solvents, and other compounds using miscible cosolvents was conducted at Operable Unit 1, Hill Air Force Base, UT. Most of the light nonaqueous phase liquid (LNAPL) consists of low solubility, low vapor pressure compounds. It is not miscible in pure methanol, ethanol, acetone, or isopropanol. The field experiment involved the injection and extraction of 7,000 gallons of a mixture of tert-butanol and n-hexanol in a 10 by 15 ft confined cell. The results of soil coring indicate better than 90% removal of the more soluble contaminants and 70% to 80% removal of less soluble compounds. The results of NAPL partitioning tracer tests show about 80% removal of the total NAPL.

A series of nine pilot scale NAPL treatability studies were recently conducted at Operable Unit 1 of Hill Air Force Base (AFB), Utah. These studies were performed under the Strategic Environmental Research and Development Program (SERDP) and the Advanced Applied Technology Demonstration Facility (AATDF). The field activities included field tests of air sparging, steam injection, in-well aeration, cyclo-dextrin flooding, surfactant mobilization, two surfactant solubilization tests, cosolvent solubilization, and cosolvent mobilization. The purpose of these studies was to evaluate and compare promising NAPL remediation technologies under similar conditions.

Operable Unit 1 (OU1) at Hill AFB was chosen as the site for the demonstrations due to the relatively large amounts of NAPL, and due to the presence of a shallow, low permeability confining layer beneath the contaminated zone. Each test was con-ducted inside a confined cell, created by installing interlocking joint sheet pile walls

102

into the confining clay layer. The study area is contaminated by a complex, weathered mixture of petroleum hydrocarbons, chlorinated solvents, and other compounds. Components of this LNAPL include undecane, decane, xylenes, toluene, trimethylbenzene, naphthalene, ethylbenzene, dichlorobenzene, trichloroethane, trichloroethylene, and benzene, but these components together only account for about 3-4% of the NAPL mass (*1*). The remainder of the NAPL composition has not been positively identified, but it probably consists of weathered petroleum components having very low solubilities and vapor pressures.

In-situ cosolvent flooding involves the subsurface injection of a water miscible solvent such as an alcohol. The miscible alcohol achieves NAPL removal through a number of complementary mechanisms including the reduction of interfacial tension (IFT) between the aqueous and NAPL phases, the enhancement of NAPL component solubilities in the aqueous phase, swelling of the NAPL phase due to alcohol partitioning, and, under certain conditions, complete miscibility of the aqueous and NAPL phases. By manipulating the cosolvent formulation it is possible to design a cosolvent flood to preferentially dissolve or mobilize NAPL, as desired.

The two strategies for NAPL recovery using cosolvents each require special design considerations. The goal in an enhanced dissolution flood is to dissolve the NAPL without mobilizing it. In contrast, the goal in a NAPL mobilizing flood is to remove the NAPL as a separate phase as rapidly as possible. Brandes and Farley (*2*) demonstrated that the dominant mode of NAPL recovery depends largely on the nature of the phase behavior of the NAPL/cosolvent/water system. Additional details on the theory of cosolvent flood design using ternary phase diagrams may be found in Falta (*3*) and Falta *et al.* (*1*).

The purpose of our field test was to mobilize and remove NAPL using miscible alcohols. OU1 at Hill AFB was the site of a previous cosolvent dissolution field experiment conducted by the University of Florida and the USEPA in the summer of 1995 (*4*). Our field experiment operated under similar site conditions to the Florida experiment, as well as to the seven other field remediation experiments that were performed at the site in the summer of 1996. The two cosolvent experiments (the University of Florida test and our test) are the first field applications of cosolvent flooding.

Cosolvent Selection

The Hill AFB NAPL has some unusual phase properties in that it does not completely dissolve in pure acetone, methanol, ethanol, or isopropanol (Table I). Our goal during the cosolvent selection process was to find a cosolvent system that would maximize the separate phase displacement of the NAPL in the field test cell. A condition for this type of displacement is that the NAPL must be completely soluble in the cosolvent, preferably at a relatively low cosolvent concentration.

Our approach to cosolvent selection was to screen candidates by their ability to form a single phase liquid with the NAPL. After a series of simple batch tests using only the cosolvent and NAPL (no water), it was found that only the higher molecular weight alcohols were capable of achieving full miscibility with the NAPL. Moreover, these alcohols only formed single-phase mixtures at high concentrations. With increasing molecular weight, the alcohols became more effective at forming single-phase mixtures with the NAPL; however, these alcohols have low aqueous solubilities

Table I. Screening Evaluation of Candidate Alcohols

Candidate Alcohol	Hill AFB OU1 LNAPL Miscible in Pure Alcohol?	Alcohol Solubility in Water (weight %)
methanol	no	infinite
ethanol	no	infinite
isopropanol (IPA)	no	infinite
tert-butyl alcohol (TBA)	yes	infinite
n-butanol	yes	9%
n-pentanol	yes	1%
n-hexanol	yes	0.6%
n-octanol	yes	0.05%

(Table I). From Table I, it is clear that of the miscible alcohols, only tert-butyl alcohol (TBA) is capable of forming a single-phase mixture with the NAPL. We also tested other miscible solvents such as various propylene glycols, but these were found to be ineffective.

From our observations that the higher molecular weight alcohols such as hexanol were more effective at forming a single phase with the Hill NAPL, we began to evalu-

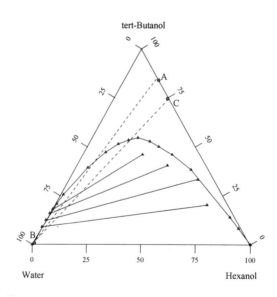

Figure 1. Ternary phase diagram for a TBA/hexanol/water system.

ate the behavior of mixtures of alcohols with the NAPL in the presence of water. Because alcohols such as hexanol have limited aqueous solubilities, it is necessary to add a cosolvent such as TBA in order to make them fully miscible. Figure 1 shows a ternary phase diagram for a system of TBA, hexanol, and water.

From the figure, it can be seen that it is necessary to maintain a TBA volume concentration of about 55% to keep the hexanol fully miscible. After constructing a number of these alcohol-alcohol-water phase diagrams for various alcohol pairs, we settled on a mixture of 85% TBA, 15% hexanol as the potential flooding solution for the field test.

A pseudo-ternary diagram (the alcohols are lumped into one component) for a TBA/hexanol, Hill NAPL, water system is shown in Figure 2. This system has a very high binodal curve, with a very small single-phase zone. The tie lines for this system (not shown here) have a very strong negative slope, indicating that the cosolvents do not partition significantly into the Hill NAPL. From fractional flow theory, (5-7, 1) it is clear that a flooding solution using these cosolvents must be injected at a very high concentration (>95%) in order to promote effective NAPL mobilization. The TBA/hexanol, NAPL, water phase diagram contains a three-phase zone, as does the

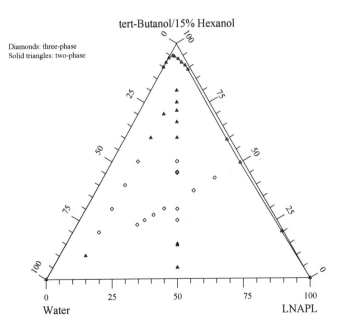

Figure 2. Ternary phase diagram for a TBA-hexanol/Hill OU1 NAPL/water system.

TBA, NAPL, water phase diagram (not shown). For the multicomponent LNAPL from Hill AFB, the formation of the third phase may be due in part to fractionation of LNAPL components. Under the anticipated flooding conditions of high cosolvent concentration, the three phase effects were not expected to be a significant problem although some three-phase samples were collected during the cosolvent flood.

A series of column tests were performed to evaluate the performance of the selected cosolvent mixtures with the Hill NAPL. Contaminated soil samples from the test cell were blended to form a homogeneous mixture, and were then packed into laboratory columns. Each column test consisted of a NAPL partitioning tracer test, followed by a cosolvent flood, finishing with a partitioning tracer test. During the cosolvent floods, the effluent concentrations of selected NAPL components (xylene, decane, undecane, dichlorobenzene, and naphthalene) were measured. However, because these compounds only represent a small fraction of the total NAPL mass it was necessary to estimate the overall flooding efficiency through the use of the partitioning tracer tests, or through Sohxlet extractions of the columns.

NAPL partitioning tracer tests involve the simultaneous injection of conservative (non-partitioning) and nonconservative (NAPL partitioning) tracers. If the equilibrium NAPL-water tracer partition coefficients (K_p) are measured in batch tests, then the volume of NAPL in the test zone can be determined from the chromatographic separation of the tracer breakthrough curves using the method of moments (8). In other words, if NAPL is present, the partitioning tracers are retarded relative to the non-partitioning tracers, and the volume of NAPL may be determined from the curve separations.

The column tests with pure TBA and with TBA/hexanol blends indicated that about 80% of the NAPL could be removed with a two to three pore volume flood. The TBA/hexanol floods were more efficient than TBA alone, and they tended to cause significant mobilization of the NAPL whereas the pure TBA did not. However, due to the low aqueous solubility of hexanol, it was necessary to use a mixture of TBA and hexanol. The various laboratory tests are summarized by Falta *et al.* (*1*), and full details are available in the works of Price (*9*), Wright (*10*), Haskell (*11*), and Meyers (*12*).

Field Test Design

The test cell was designed to hydraulically confine the pilot test in order to contain the flooding solution as well as any mobilized NAPL. The test cell has horizontal dimensions of about 15 ft by 10 ft, and the cell walls consist of interlocking sheet pile walls. These sheet piles were driven about 6 ft into the underlying clay, and the sheet pile joints were grouted. Side and top views of the test cell are shown in Figures 3 and 4, respectively. Based on hydraulic and tracer testing, the 12-foot thick treatment zone in the test cell has a pore volume of 1800 - 1900 gallons.

As shown in Figure 4, the cell includes four injection wells and three extraction wells, located in parallel lines at each end of the cell. One of the injection wells, U1-2342, was offset due to the loss of an auger flight during drilling, and was not used on the subsequent tests. Eleven bundles of multilevel samplers were located in the flooding area on the interior part of the cell, and each bundle consisted of five evenly

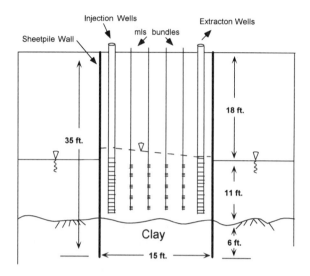

Figure 3. Side view of test cell configuration.

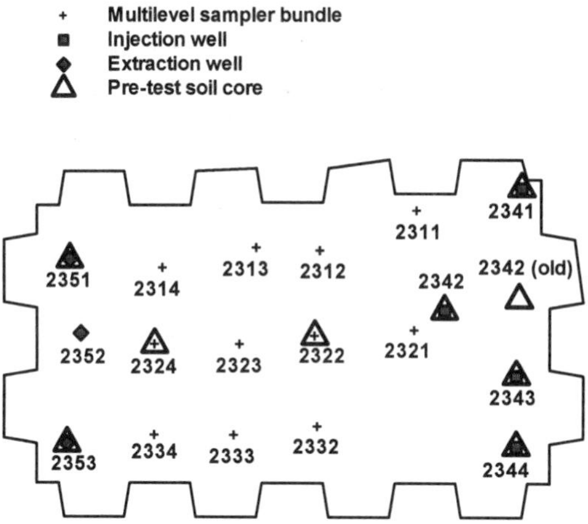

+   **Multilevel sampler bundle**
■   **Injection well**
♦   **Extraction well**
△   **Pre-test soil core**

Figure 4. Top view of test cell showing wells, samplers, and pre-flood soil sampling locations. The extraction wells are on the north side of the cell.

spaced samplers (see Figure 3). These were constructed with stainless steel tubing with 40-micron filters, and were installed using the EPA cone penetrometer.

Figure 5 shows a schematic of the site layout developed for the test. Clean water was piped in from a Hill AFB potable water supply, and stored in a 6000 gallon tank with a float valve to maintain the tank at a constant level. A 500 gallon tracer tank was attached to the line coming out of the water tank, with valves to allow switching between the tanks. The cosolvent solutions were stored in two 6,000-gallon tanks located on an HDPE liner, some distance from the test cell. The inlet line was run to a pump house, where the line was split into three injection well lines, one each for wells U1-2341, U1-2343, and U1-2344. Two Cole-Parmer brand industrial/process variable speed peristaltic pumps were used for the fluid injection. A series of flow-meters and sampling tubes were installed just downstream of the pumps. The sampling tubes were plumbed so that all of the flow for each well tube could be diverted through the corresponding tube. These sampling tubes allowed for precise volumetric measurements of the flow rates, and they were used to calibrate digital pump drive controllers. The injection well lines leaving the pump house were made of ¾-inch i.d. reinforced flexible PVC tubing. The tubing was run from the pump house to the bottom of each injection well (a depth of about 27 feet). For the partitioning tracer tests, the bottom end of each tube was sealed, and holes were drilled in the bottom ten feet of each injection well tube, so that the fluids would be uniformly injected over the injection well screen zones. For the cosolvent flood, fluid was only injected into the top three feet of the screen zone, and the remainder was packed off with inflatable packers to prevent loss of cosolvent to a high permeability zone.

The three extraction wells (U1-2351, U1-2352, and U1-2353) were plumbed in a similar manner. Reinforced flexible PVC tubing was extended from the bottom of the wells to the pump house. For the tracer tests, the bottom of each tube was sealed, and holes were drilled in the bottom ten feet of the tubes (eight feet for well U1-2352, which was only 25 feet deep). During the cosolvent flood, only the top 3-4 feet of the screened zone was used, and the remainder was packed off. At the pump house, the PVC tubing was connected to the peristaltic pump tubing. Two of the peristaltic pumps described above were used for the extraction wells. Peristaltic pumps were chosen for this application because the pumped fluid is never in contact with the motor or other electrical equipment. This is an important consideration when dealing with flammable liquids such as alcohols. Since the water table was located at a depth of about 17 feet in the test cell, it is possible to use a suction lift to deliver the fluids to the surface. The maximum suction for the peristaltic pumps was about 24 to 25 feet at the site (located at an elevation of nearly 5000 feet above sea level). The flow meters and sampling tubes were installed downstream of the pumps. The sampling tubes were used for fluid sampling during the tests, and for calibrating the pump drives.

From the pump station, the three extraction lines were combined into a single line. This line was run to three tanks located on a high-density polyethylene (HDPE) liner. These tanks included two 21,000-gallon tanks and a 6,000-gallon tank. One of the 21,000-gallon tanks was used for dilute wastes, and was connected to the Base industrial water treatment plant. The other two tanks were used for onsite storage of concentrated wastes, which were later incinerated. Unlike the injection side, which

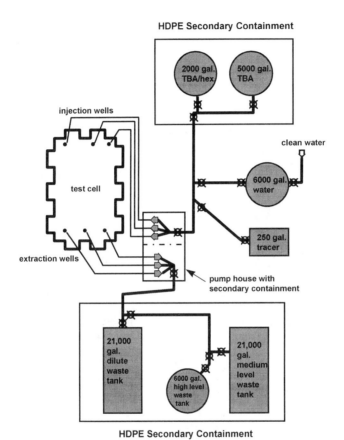

Figure 5. Site layout for cosolvent flooding test.

could operate under a gravity feed, the extraction side of the system was subjected to fairly high pumping loads, including a suction lift of about 20 feet (the pumps were about three feet off the ground), and an outlet head of about 10 feet to reach the top of the waste tanks, where the fluids were discharged. While the pump motors were easily capable of handling this load, the peristaltic tubing on the extraction well pumps required periodic replacement, and frequent adjustment and maintenance to keep a steady flow rate. Additional details on the site design and experimental procedures are available in Hill AFB (*13*) and Falta *et al.* (*1*).

Test Results

The pilot test consisted of three main parts: pre-flood sampling and characterization, cosolvent flooding, and post-flood sampling and characterization. The pre-test sampling consisted of soil sampling and a partitioning tracer test. Soil samples were taken from about eight vertical locations in eight soil borings shown in Figure 4 and analyzed for selected NAPL compounds. The partitioning tracer test consisted of the injection of a little over 0.1 pore volumes of tracer solution followed by several pore volumes of water. Tracer breakthrough curves were measured in the extraction wells, and in the multilevel samplers.

The pre-flood characterization work indicated that the NAPL was mainly present in the upper 6 feet of the treatment zone. Figure 6 shows a cross sectional view of the NAPL saturation in the test cell (perpendicular to the direction of flow) determined from the multilevel sampler breakthrough curves. The pre-flood soils data indicate a similar profile with much higher concentrations in the upper 6 feet of the saturated zone. The initial characterization also revealed the presence of a very high permeability zone in the bottom part of the test cell, where there was little contamination. The bottom 2/3 of the injection and extraction wells were packed off during the cosolvent flood in order to avoid cosolvent losses to this zone. The cosolvent flood involved several parts. During the first 1300 gallons of injection, 95% TBA was injected. The flooding solution was then switched over to an 80% TBA, 15% hexanol, 5% water solution for 2000 gallons. The hexanol/TBA blend was followed by more than 4000 gallons of 95% TBA, finishing with a long water flood to remove the cosolvents.

The 7000 gallons of tert-butanol used in the test was a gasoline grade product donated by the Arco Chemical Company. Gasoline grade TBA has a purity of at least 96%, with small amounts of water and methanol. The total cosolvent flooding volume was approximately 4 pore volumes. This volume is larger than would normally be expected for a mobilizing cosolvent flood, but the added volume was needed due to the unfavorable phase characteristics of the Hill NAPL as shown in Figure 2.

The effluent from the test cell was routed to one of the three tanks depending on the alcohol concentrations. Very concentrated effluent (>90% alcohol) was discharged to the 6000 gallon tank, to be used as an incinerator fuel. Dilute wastes from the tracer test and water floods were temporarily stored in a 21,000 gallon tank, and then pumped to the Hill AFB industrial waste water treatment facility. Intermediate wastes were discharged to a 21,000-gallon tank and then sent to an incinerator.

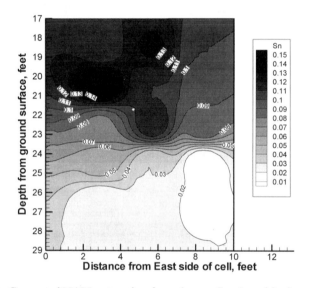

Figure 6. Computed NAPL saturation from the pre-flood partitioning tracer test.

Significant amounts of NAPL were observed in the extraction well effluent lines and the multilevel samplers during the flood. Both the samplers and the wells showed a similar pattern as the cosolvent progressed through the cell. First, the samples would become cloudy as the alcohol reached the sampling point. Next, various amounts of mobilized NAPL would appear in the solution, which became highly discolored. As the flood progressed, the NAPL no longer appeared in samples, and the solution became nearly clear again, except in a few of the sampler locations near the extraction end of the cell which remained discolored throughout the test.

After the cosolvent was flushed from the cell, a final partitioning tracer test was performed using the same methods as the initial tracer test. Soil samples were taken from eight vertical locations in six soil borings and analyzed for the same NAPL compounds as in the initial soil sampling. Figure 7 shows the computed NAPL saturation in the test cell (a cross-section perpendicular to the flow) from the post-flood multilevel sampler tracer data. A comparison of the NAPL distributions used in Figures 6 and 7 indicate a reduction of NAPL saturation of 80%.

Table II summarizes the results of moment analyses of tracer data from the extraction wells. The extraction wells indicate an overall NAPL removal of about 78% which is similar to the multilevel sampler results. A higher removal efficiency was noted along the center line of the test cell. The tracer tests showed that this zone was swept more effectively, so more of the cosolvent passed through this region. The fact that this zone had a higher fraction of removal indicates that a larger volume of cosolvent would have produced a better NAPL removal from the test cell overall.

Table II. Comparison of pre-flood and post-flood NAPL saturation estimates using the bromide/2,2-dimethyl-3-pentanol extraction well data from both tests.

	Pre-Flood NAPL Volume (gallons)	Post-Flood NAPL Volume (gallons)	Pre-Flood NAPL saturation	Post-Flood NAPL saturation	Percent Reduction in NAPL saturation
Well U1-2351	40.3	12.5	0.065	0.018	72.3%
Well U1-2352	41.2	6.0	0.085	0.011	87.1%
Well U1-2353	43.0	10.4	0.059	0.015	74.6%
Overall	124.5	28.9	0.068	0.015	77.9%

The soils data are summarized in Table III. These averages were compiled using all of the soil samples, 63 before the flood, and 60 after the flood. Again the overall removal is about 78%, with much higher removal rates for the more soluble compounds. If NAPL mobilization were the only mechanism of NAPL removal, the efficiencies would be expected to be similar for all compounds. The higher removal efficiency for the lighter compounds indicates that NAPL dissolution was also significant during the experiment.

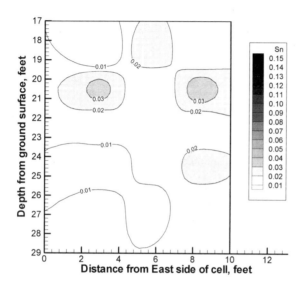

Figure 7. Computed NAPL saturation from the post-flood partitioning tracer test.

Table III. Comparison of pre-flood and post-flood average soil mass fractions of target compounds. These arithmetic averages are based on 63 pre-flood samples and 60 post-flood samples.

Chemical	Pre-Flood Average Soil Mass Fraction (mg/kg)	Post-Flood Average Soil Mass Fraction (mg/kg)	Percent Reduction in Average Soil Mass Fraction
undecane	74.47	14.80	80.1%
decane	35.34	10.18	71.2%
1,3,5-trimethylbenzene	5.31	0.376	92.9%
o-xylene	3.27	0.230	93.0%
naphthalene	1.93	0.251	87.0%
1,2-dichlorobenzene	1.48	1.431*	3.3%*
m-xylene	1.22	0.115	90.6%
toluene	2.20	0.139	93.7%
1,1,1-trichloroethane	0.47	0.019	96.0%
ethylbenzene	0.60	0.048	92.0%
trichloroethylene	0.015**	0.007**	53.0%**
benzene	0.004**	0.009**	-125%**
Sum of Target Chemicals	**126.3**	**27.6**	**78.1%**

*Two of the post-test core samples contained 1,2-dichlorobenzene mass fractions which were nearly an order of magnitude higher that the highest pre-flood 1,2-dichlorobenzene mass fraction.
**The mass fractions of these compounds were close to or below detection limits for most samples.

Figures 8 and 9 show the pre-flood and post-flood profiles of undecane and o-xylene soil mass fraction. Most of the other compounds show a similar pattern. In each case, there is a one to two order of magnitude reduction in the soil concentration except for a few post-flood samples that remain high. The post-flood averages shown in Table III reflect these few high remaining concentrations. If the median values of concentration are compared before and after the flood, the average reduction is from 95% to 100% for all chemicals. Comparing the individual soil borings, it was found that the high concentrations in the post-flood data all came from the extraction end of the test cell. Because the initial distribution of contaminants was relatively uniform along the test cell, this is further evidence that a larger volume of cosolvent would have removed more of the contaminants from the cell.

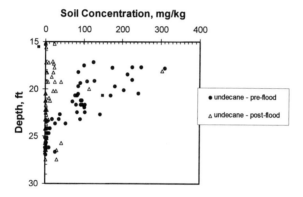

Figure 8. Comparison of pre-flood and post-flood soil mass fractions of undecane.

116

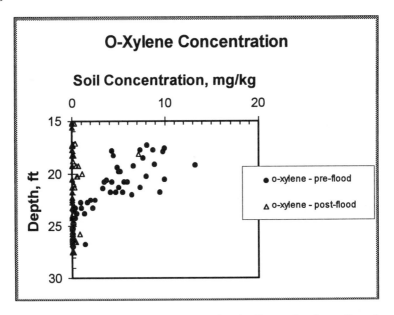

Figure 9. Comparison of pre-flood and post-flood soil mass fractions of o-xylene.

Acknowledgments

The authors gratefully acknowledge the help of Charles Wright, Sarah Price, John Coates, Patrick Haskell, Julie Olivier, Katherine Zitch, Rex Hodges, Brett Whiten, Ron Anderson, and Robert Cullom from Clemson University, and Sally Benson, Ray Solbau, April James, and John Clyde from the Berkeley National Laboratory. This project would not have been possible without their efforts in the field and laboratory. We would like to thank Arco Chemical Company for donating the 7000 gallons of gasoline grade tert-butanol used in the experiment. The authors would also like to thank Hill Air Force Base and the Regional State and EPA regulators for their cooperation. This study was supported by the USEPA Robert S. Kerr Environmental Research Laboratory under Cooperative Agreement CR-821992-01-0. The information in this document has been funded in part by the U.S. Environmental Protection Agency under Cooperative Agreement No. CR-821992 to Clemson University with funds provided through the Strategic Environmental Research and Development Program (SERDP). This document has not been subjected to Agency review; mention of trade names or commercial products does not constitute endorsement or recommendation for use.

Literature Cited

1. Falta, R.W.; Lee, C.M.; Brame, S.E.; Roeder, E.; Wright, C.L.; Price, S.W.; Coates, J.T.; Haskell, P.A.; Wood, A.L.; Enfield, C.G., *Field Evaluation of*

Cosolvent-Enhanced In-Situ Remediation, Draft Final Report for Cooperative Agreement CR-821992-01-0, US EPA, R.S. Kerr Environmental Research Lab., 1997.

2. Brandes, D.; Farley, K. *Water Environment Research*, **1993**, 65, 869-878.
3. Falta, R.W. *Ground Water Monitoring and Remediation*, in press, **1997**.
4. Annable, M.D.; Rao, P.S.C.; Sillan, R.K.; Hatfield, K.; Graham, W.D.; Wood, A.L.; Enfield, C.G. In *Non-Aqueous Phase Liquids (NAPLS) in Subsurface Environment, Proceedings of the specialty conference held in conjunction with the ASCE National Convention*, Reddi, L N., Ed.; ASCE:Washington, D.C., 1996, 212-220.
5. Larson, R.G.; Hirasaki, G. *Soc. Pet. Eng. J.*, **1978**, February, 42-58.
6. Pope, G.A. *Soc. Pet. Eng. J.*, **1980**, June, 191-205.
7. Lake, L.W., *Enhanced Oil Recovery*, Prentice Hall, Inc.:Englewood Cliffs, NJ, 1989.
8. Jin, M..; Delshad, M.; Dwarakanath, V.; McKinney, D.C.; Pope, G.A.; Sepejrnoori, K.; Tilburg, C.; Jackson, R.E. *Water Resources Research*, **1995**, 31, 1201-1211.
9. Price, S.W. Selection of a Cosolvent Flooding Solution for Removal of a Multicomponent LNAPL at Operable Unit One, Hill Air Force Base, Utah. **1997**, Clemson University. M.S. thesis.
10. Wright, C. L. The Effect of Cosolvent Flooding on NAPL/Water Partition Coefficients. **1996**, Clemson University. M.S. thesis.
11. Haskell, P.A. Use of Surrogate Compounds to Monitor the Removal of a Non-Aqueous Phase Liquid during a Cosolvent Flood. **1997**, Clemson University. M.S. thesis.
12. Meyers, S.L. The Effect of Cosolvent Flooding on NAPL Composition and Partition Coefficients of Tracers used in Partitioning Interwell Tracer Tests. **1997**, Clemson University. M.S. thesis.
13. Hill AFB, *Phase I Work Plan for Eight Treatability Studies at Operable Unit 1*, August, 1995.

Chapter 9

Field Test of Cyclodextrin for Enhanced In-Situ Flushing of Multiple-Component Immiscible Organic Liquid Contamination: Project Overview and Initial Results

Mark L. Brusseau[1,2], John E. McCray[2], Gwynn R. Johnson[1], Xiaojiang Wang[1], A. Lynn Wood[3], and Carl Enfield[3]

[1]Departments of Soil, Water, and Environmental Science and [2]Hydrology and Water Resources, University of Arizona, Tucson, AZ 85721
[3]National Risk Management Research Laboratory/Subsurface Remediation Division at Ada, U.S. Environmental Protection Agency, 911 Kerr Research Drive, Ada, OK 74820

The purpose of this paper is to present an overview and the initial results of a pilot-scale experiment designed to test the use of cyclodextrin for enhanced in-situ flushing of an aquifer contaminated by immiscible liquid. This is the first field test of this technology, termed a complexing sugar flush (CSF). The field test was conducted within a solvent and fuel disposal site at Hill Air Force Base, UT. The cyclodextrin solution increased the aqueous concentrations of all the target contaminants to values from about 100 to more than 20000 times the concentrations obtained during the water flush conducted prior to the CSF. Concomitantly, the CSF greatly enhanced the rate of mass removal during the 8 pore-volume flush, which resulted in a 41% reduction in contaminant mass. Based on these results, it is clear that the CSF technology was successful in enhancing the remediation of the immiscible-liquid contaminated site. There are several attributes of cyclodextrin that in some situations may offer advantages compared to using surfactants or cosolvents for solubilization-based enhanced flushing.

The limitations of current technologies for cleaning up sites contaminated by immiscible liquids are well known (e.g., 1). This has lead to a search for alternative technologies. One approach that is receiving serious consideration involves the use of a flushing solution containing a chemical agent that can greatly increase the mass of contaminant held in the aqueous phase. These enhanced solubilization agents include surfactants, cosolvents, and complexing agents (e.g., dissolved organic matter,

118

cyclodextrin). Surfactants, and cosolvents to a lesser extent, have received the majority of attention to date. While these agents have several advantageous properties, and have been used successfully for enhanced flushing, they also have some disadvantages. Thus, investigation of alternative agents continues.

The purpose of this paper is to present an overview and the initial results of a pilot-scale experiment designed to test the use of cyclodextrin for enhanced in-situ flushing of an aquifer contaminated by immiscible organic liquid. The field test was conducted within a site at Hill Air Force Base, Utah, that is contaminated with a complex, multiple-component, immiscible organic liquid comprised of fuels and spent solvents. The study was conducted as part of a project organized by the Environmental Protection Agency National Research Laboratory, Ada, OK, which was funded by the SERDP Program of the U.S. Department of Defense. The purpose of the project was to evaluate the performance of several innovative remediation technologies, implemented simultaneously at the same site (2).

BACKGROUND
Basic Properties of Cyclodextrin

The compound used in the experiment was hydroxypropyl-β-cyclodextrin (HPCD), which consists primarily of glucose molecules. Cyclodextrins are cyclic oligosaccharides produced by bacterial biodegradation of starch. The three major types of base cyclodextrins, α, β, and γ, are composed of six, seven, and eight glucose molecules, respectively. The glucose molecules are arranged in a toroidal shape as illustrated in Figure 1. The external surface of the compound is polar, while the interior is apolar. An important property of cyclodextrins is their ability to "complex" a wide variety of solutes, thereby increasing their apparent aqueous solubilities. The enhanced solubilization occurs by inclusion of the solute into the apolar cavity of the cyclodextrin. This property has resulted in the use of cyclodextrins for pharmaceutical applications (e.g., drug delivery), and is the basis for the use of cyclodextrins for enhanced flushing of contaminated soils and aquifers, as will be discussed below.

The molecular weights of the base cyclodextrins range from 972 to 1297. The molecular weight of HPCD ranges from 1326 to 1500, depending on the degree of hydroxypropyl substitution. The outer diameter of a cyclodextrin molecule is about 1.5 nm. The pK_a of the hydroxyl groups associated with the external surface of cyclodextrin is about 12 (3). Thus, cyclodextrins will remain polar in essentially all environmental aqueous systems. The aqueous solubilities of the base cyclodextrins range from 18 to 230 g/l (3). Their solubilities can be increased by substitution of polar functional groups. For example, hydroxypropyl-β-cyclodextrin (HPCD), the focus of our work, has a solubility greater than 500 g/l. These high solubilities allow relatively high concentrations to be used in field applications.

Given the composition of cyclodextrins, they are essentially non-toxic to humans. Thus, they should be readily acceptable for injection into subsurface systems, including sole-source aquifers. Cyclodextrins also appear to be nontoxic to microorganisms. For example, the presence of high concentrations (up to 10%) of a specific cyclodextrin (HPCD) actually enhanced the magnitude of biodegradation of an organic compound, indicating no apparent negative impact of the cyclodextrin on the microbial population (4).

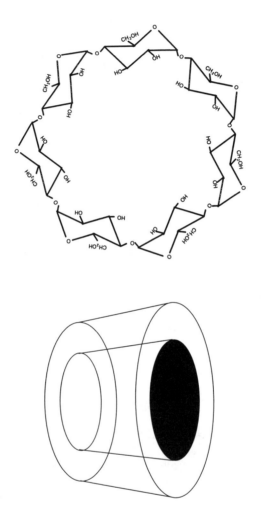

Figure 1.   Schematic of a cyclodextrin molecule.

Behavior in Environmental Systems

Based on the external surface characteristics of cyclodextrin, one would expect low reactivity between cyclodextrins and porous media. Furthermore, given their small size, it is unlikely that cyclodextrins will experience pore exclusion during transport in most porous media. The potential impact of sorption and pore exclusion on cyclodextrin transport was investigated by Brusseau et al. (5). Breakthrough curves obtained for transport of HPCD through a column packed with the well-known Borden aquifer material are shown in Figure 2a. Comparison of the HPCD breakthrough curves to those of pentafluorobenzoate (PFBA), a well-characterized nonreactive tracer, indicates essentially identical transport behavior. Similar results were obtained for transport in a column packed with a high organic-carbon content (12.6%) soil (see Figure 2b).

The measured retardation factors for HPCD were unity for both porous media, indicating no sorption. The absence of HPCD sorption was substantiated by the results of batch experiments. Additional experiments have shown minimal or no sorption of HPCD by glass beads or kaolinite (6), or by two sandy aquifer materials (unpublished data). These results support the absence of pore exclusion as well as that of sorption. Thus, for advection-dominated systems, cyclodextrin should exhibit transport behavior identical to nonreactive tracers. This was confirmed by the results obtained from tracer tests conducted at the Hill AFB site, as discussed by Brusseau et al. (7).

Cyclodextrins are stable under typical environmental conditions. For example, the half-life for hydrolysis of cyclodextrin is 48 days under extreme conditions (pH < 0, T = 40 ^{0}C) (3) and would be much longer under environmental conditions. Cyclodextrins are very resistant to precipitation. For example, the addition of 10 g/l of $CaCl_2$ to a solution of carboxymethyl-β-cyclodextrin did not cause precipitation (8). Given the composition of cyclodextrin, it is expected to be biodegradable over the long term. However, HPCD has been shown to be resistant to biodegradation for time periods of at least a few months (4). Based on these properties, cyclodextrin can be expected to be mass conservative during typical field applications.

The presence of cyclodextrin in water has negligible impact on pH and ionic strength. The addition of cyclodextrin to water causes a slight reduction in surface tension [reduced from 72.5 to approximately 62 dynes/cm for a 10% HPCD solution] (9). A 10% HPCD solution has a density of 1.025 g/mL and a viscosity of 1153 uPa·s at 21.5 ∘C, which are not greatly different from those for pure water [1.00 and 971, respectively]. Thus, the addition of cyclodextrin should not alter the physical/chemical properties of the solution.

Cyclodextrin does not significantly reduce the interfacial tension of the immiscible liquid-water interface and, therefore, is unlikely to mobilize immiscible liquid saturation (10,11). For example, the interfacial tension for a sample of NAPL collected from the site was reduced from 35 to 10 dynes/cm in a 10% HPCD solution. While mobilization can enhance removal of immiscible liquids from the subsurface, it can be difficult, in many instances, to capture all mobilized immiscible liquid during remediation. Thus, remediation techniques based on mobilization may not be appropriate under many circumstances.

122

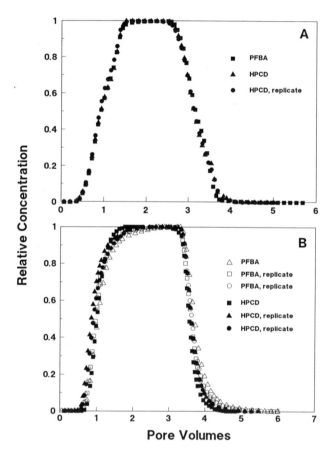

Figure 2.   Breakthrough   curves   for   transport   of   cyclodextrin and
pentafluorobenzoate through columns packed with porous media; A)
Borden aquifer material; B) High organic-carbon content soil. Figure
from Brusseau et al. (4).

Enhanced Solubilization and Mass Removal

Enhanced solubilization of low solubility organic compounds into the aqueous phase is obtained by the effect of the cyclodextrin on the aqueous activity of the compound. The apparent solubilities of single organic compounds in aqueous solutions containing cyclodextrin have been observed to increase linearly with the concentration of cyclodextrin. A simple relationship describing the relationship between cyclodextrin concentration (C_H) and the solubility (S^w) of a single solute is (9):

$$S^A = S^w (1 + K_{cw} C_H) = S^w E \qquad (1)$$

where S^A is the apparent solubility of the organic solute in the cyclodextrin solution, K_{cw} is the partition coefficient of the solute between cyclodextrin and water, and E is the solubility enhancement factor.

The impact of cyclodextrin on apparent solubility is greater for more hydrophobic compounds, similar to the effect of surfactants and cosolvents. This behavior is illustrated in Figure 3, wherein is presented a plot of log K_{cw} versus log K_{ow}. Clearly, the magnitude of the apparent solubility enhancement increases with increasing K_{ow} of the compound. Thus, a compound such as anthracene will experience a far greater relative solubility enhancement compared to a more soluble compound such as trichloroethene.

It should be noted that the greater relative solubility enhancement observed for more hydrophobic compounds does not necessarily mean that the rate of mass removal of a more hydrophobic compound from a multi-component system would be greater than that of a less hydrophobic compound. The rate of mass removal is a direct function of the total mass of contaminant (dissolved + complexed) in the solution being flushed through the porous medium. While the apparent solubility of a more hydrophobic compound is enhanced greatly relative to its aqueous solubility, the total mass of the contaminant in solution may still be smaller than that of a less hydrophobic compound. This can be illustrated for a system with trichloroethene and anthracene in solutions containing 10% HPCD. For this system, the apparent solubility of trichloroethene is increased by 6 times, whereas that of anthracene is increased by a factor of 293. The total concentration (i.e., apparent solubility) of trichloroethene in a 10% HPCD solution is approximately 6,700 mg/L, whereas that of anthracene is 14 mg/L. Thus, for a multi-component immiscible liquid wherein the two compounds comprise equal mole fractions, the mass-removal rate for trichloroethene will be almost 500 times greater than that for anthracene, assuming no differential effects of nonideal dissolution factors. This example indicates the importance of recognizing the difference between the magnitude of enhanced solubility and the resultant impact on mass removal.

COMPLEXING SUGAR FLUSH PILOT TEST
Contaminant Location and Distribution

Several waste disposal areas are located within the site (operable unit 1) wherein the tests were conducted (2). Two chemical disposal pits were in operation at OU1 from 1952-1973. Large quantities of liquid wastes (mainly waste fuels, and spent solvents) were disposed of and periodically burned in the pits. Inactive fire training areas are also present at the site. These fire training areas were used by Hill AFB until 1973 as a

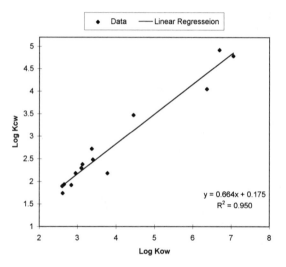

Figure 3.   Correlation between cyclodextrin-water partition coefficient (K_{cw}) and octanol-water partition coefficient. Figure from Wang and Brusseau (8).

practice area to extinguish simulated aircraft fires. Fuels burned in this area included jet fuel, oil, and combustible waste chemicals. A landfill, used for industrial liquid and solid waste disposal, was in operation from 1940 to 1978 and contains industrial sludge, waste solvents, and unidentified chemicals. Another landfill, used for domestic and construction refuse, was in operation from 1967 to 1973. Finally, a brick-lined waste phenol/oil pit was used periodically from 1954-1965 to burn waste oil and phenol.

The treatment cell used for our experiment is located within or adjacent to one of the chemical disposal pits. The cell was emplaced in what is considered to be a source area of water-immiscible organic contaminants (free phase or residual saturation present). The immiscible liquid is comprised primarily of petroleum hydrocarbons, chlorinated hydrocarbons, and PAHs, and is considered to be less dense than water. The immiscible liquid is smeared throughout the saturated, as well as unsaturated, portions of the aquifer as a result of water table fluctuations. The initial immiscible-liquid saturation was estimated to be approximately 12.6% based on results from a partitioning tracer study (12).

Prior to the experiment, 12 "target contaminants" were selected from the many compounds in the immiscible liquid for the purpose of evaluating remediation effectiveness. These targets are: trichloroethene, 1,1,1-trichloroethane, benzene, toluene, o-xylene, m,p-xylene (measured as p-xylene), ethylbenzene, 1,2,4-trimethylbenzene, 1,2-dichlorobenzene, naphthalene, decane, and undecane. These compounds were selected to provide a representative subset of the immiscible-liquid constituents of concern present within the study area. We estimate that these compounds comprise slightly less than 10% of the total immiscible liquid within the cell. The remainder appears to be comprised primarily of higher molecular weight jet fuel components and relatively insoluble, pitch-like components.

Experimental Design

The remediation technology tested in this study is termed a "Complexing Sugar Flush" (CSF). The CSF is an enhanced-solubilization technique whereby cyclodextrin is used to foster the solubilization of immiscible liquids. The reagent was mixed with water and pumped into the contaminated zone using a horizontal flow field. The effluent was collected and subjected to selected treatment and disposal operations. An enclosed cell was used for this experiment to minimize migration of normally sparingly-soluble contaminants that could experience enhanced solubilization and transport in the presence of cyclodextrin, and to facilitate mass-balance analysis and performance assessment. The 3 m by 5 m area cell was enclosed by sealable 9.5 mm sheet pile walls, which were driven into the low-permeability clay layer. The top of the clay layer is approximately 8-9 m below ground surface (BGS) in the treatment area, although this depth varies with location.

A line of four injection wells and a line of three extraction wells, both normal to the direction of flow (which was approximately from south to north), were used to generate a steady-state flow field (Figure 4). The injection and extraction wells (5.1 cm diameter) were fully screened over the saturated thickness. Injection and extraction flow was generated with peristaltic pumps (Master Flex I/P with bench-top controller, Tygon LFL tubing from Cole-Parmer) using a separate pump head for each well. Technical-grade cyclodextrin was used for the experiment, and was delivered in

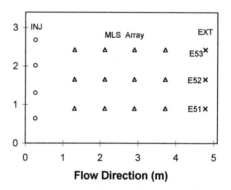

Figure 4.   Schematic of test cell.

dehydrated form (Cerestar USA Inc., Lot 8028). This technical-grade product was comprised of approximately 90% hydroxypropyl-β-cyclodextrin (HPCD), and 10% production byproducts (primarily hydrated ash). The cyclodextrin was then mixed with potable water to a achieve a HPCD concentration of 10.4% (by mass) using an 80,000 L mixing tank with four 5-Hp top-mounted mixers.

Approximately 8 pore volumes of the 10% cyclodextrin solution (approximately 65400 L total) were pumped through the cell at a rate of approximately 0.8 pore volume per day, or 4.54 L/min. Tubing degradation in the peristaltic pump system caused flow rate variations of up to 10%; however, variations of greater than 5% rarely occurred for more than two hours. The flow rate was monitored and adjusted to maintain the water table in the enclosed cell at 5.4 to 5.6 m BGS. After the eighth day, flow was ceased for one day to investigate the potential for rate-limited dissolution. Two more days of cyclodextrin flushing was conducted after the flow interruption period. At the end of the experiment, approximately 5 pore volumes of cyclodextrin-free water was flushed through the cell to remove the cyclodextrin.

Aqueous samples were collected at each extraction well to monitor for target-contaminant and cyclodextrin concentrations. Samples were collected hourly for the first 36 hours of the experiment, and every 3 to 4 hours, thereafter. In addition, 12 multi-level sampling (MLS) devices were used to collect samples at five depths within the saturated zone. The MLS system, which was constructed entirely of stainless steel, was connected to a vacuum extraction system for sample collection. The sampling interval was similar to that used for the extraction wells.

Methods for Evaluation of Remediation Performance

The effectiveness of the CSF was evaluated by five methods: (1) analysis of the solubility enhancement obtained during the CSF; (2) comparison of aqueous-phase contaminant concentrations measured for groundwater samples collected under static conditions before and after remediation; (3) comparison of soil-phase contaminant concentrations measured for core samples collected before and after remediation; (4) comparison of immiscible liquid saturations (S_n) obtained from partitioning tracer tests conducted before and after remediation; and (5) comparison of contaminant mass removed during the CSF to mass removed during a water-only flush (pump-and-treat analogy) conducted prior to the CSF. This report will focus on the results associated with methods 1 and 2. Analysis of the results obtained for measures 3 and 4 are reported by McCray and Brusseau (11), and a comparison of the CSF to the water flush is reported in McCray et al. (13).

Groundwater samples were collected under no-flow (static) conditions before and after the remediation effort. Samples were collected from all water-producing MLS points located at the 5.7 m, 6.9 m, and 8.1 m BGS levels, and from selected points at the 6.3 m BGS level when samples could not be drawn from the 5.7 m-deep points. This resulted in a total of 37 sampling points. The pre-remediation samples were collected after a period for which there had been no flow for several weeks. The post-remediation samples were collected approximately 60 hours after stopping flow at the end of the study.

Analytical Methods

Cyclodextrin can be analyzed by gas chromatography after conversion to a volatile dimethylsilyl ether form (14). A simpler method involves the use of a fluorescent dye, 2-p-toluidinylnaphthalene-6-sulphonate (TNS) that is added to the samples prior to analysis (15). Cyclodextrin has no fluorescent response, and that of TNS is minimal. However, the TNS-cyclodextrin complex exhibits significant fluorescence. The detection limit for HPCD using this method is approximately 1 mg/l. For environmental systems, the presence of high concentrations of cations such as Ca^{2+} and Mg^{2+} can cause interference by interacting with the TNS. This effect, which was minimal for our system, can be minimized by addition of EDTA to the solution.

The aqueous samples for the target compounds were collected with no head space in 40 mL (extraction wells) or 8 mL (MLS points) glass vials with teflon-lined caps. These samples were delivered to the University of Arizona and stored in the dark at 4 ^{0}C until analysis. For analysis, 5 mL-portions of the samples were transferred to 20 mL glass headspace vials (Teflon-lined septum and crimp cap), allowed to equilibrate to room temperature, and analyzed by GC-FID (Shimadzu, GC-17A), fitted with a capillary column and a head-space autosampler (Tekmar, model 7000). Analytical difficulties (co-elution of unknowns) occurred for benzene and trichloroethane in samples containing cyclodextrin. Therefore, the CSF concentrations for these two compounds are considered to be unreliable and are not reported. In addition, concentrations of decane and undecane were below detection limits for all samples not containing cyclodextrin.

RESULTS

Cyclodextrin Transport and Contaminant Elution

The cyclodextrin concentrations in the effluent of the center extraction well during the CSF are shown in Figure 5. The front of the breakthrough curve arrived at approximately 1.2 day, which is equivalent to one pore volume of flushing. This indicates that the transport of cyclodextrin was not retarded. Mass balance calculations indicate that cyclodextrin transport was conservative (100.4 % recovery). Similar results were observed for the other two extraction wells. The absence of retardation and mass loss observed herein is consistent with the results of laboratory work (5).

The elution curves for dichlorobenzene for each extraction well are presented in Figure 6. The behavior exhibited by dichlorobenzene represents that observed for most of the target contaminants. The effluent concentrations of the contaminants exhibit a large, initial increase followed by a decrease to a somewhat constant value for most contaminants (see Figure 6). The latter behavior is believed to be due primarily to the effect of a decreasing mole fraction on dissolution (11).

For wells E52 and E53, the peak dichlorobenzene concentration occurred at about one day, which is coincident with the time that the peak cyclodextrin concentration (10%) arrived at these wells. The initial peak of dichlorobenzene in well E51 occurred at about 1.5 days, also concurrent with the breakthrough of 10% cyclodextrin. The later breakthrough at this well was due to the presence of lower-permeability media along the east side of the cell, as characterized by the results of tracer tests. The concurrent arrival of the contaminant peaks with the cyclodextrin

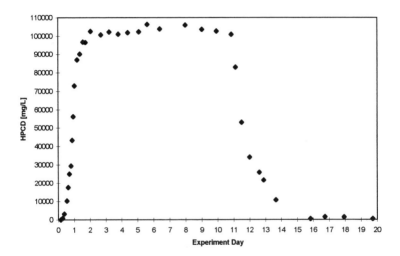

Figure 5.   Composite breakthrough curve for cyclodextrin measured for the extraction wells during CSF field test.

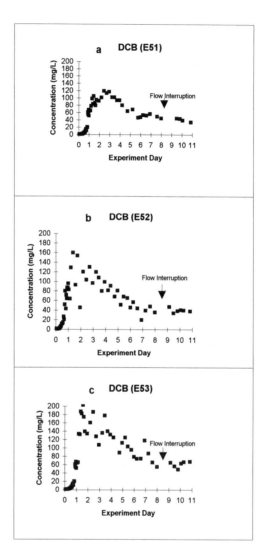

Figure 6.   Elution curves for dichlorobenzene measured at the extraction wells during the CSF field test.

breakthrough is expected based on the principles of enhanced flushing. The arrival of the contaminant elution peaks at the equivalent of approximately one pore volume is expected given that transport of cyclodextrin is not retarded.

Cyclodextrin induced very large increases in the aqueous concentrations of all the target contaminants, ranging from about 100 to more than 20000 times the concentrations associated with the water flush conducted immediately prior to the CSF (see Table 1). These large solubility enhancements produced large increases in the mass-removal rate for the target compounds. As discussed in detail by McCray and Brusseau (11), this resulted in a 41% reduction in soil-phase contaminant mass. This mass removal corresponds well to the results of the partitioning-tracer tests, which indicated a 44% reduction in average NAPL saturation (11).

Static Groundwater Concentrations

The average aqueous concentrations for the target contaminants in the samples collected at the various MLS depths are reported in Table 2. During the CSF test, some of the sampling points at the 5.7 m depth became inoperable, thus there are fewer post-remediation samples at this depth. The initial (and final) concentrations are well below the single-component solubilities, which is to be expected for a multicomponent immiscible liquid. In general, the higher solubility compounds with relatively higher mole fractions have larger concentrations, as expected.

As a result of the CSF, the aqueous concentrations are significantly reduced at the lower two MLS depths, which are at the center and at the bottom of the remediation zone. The concentrations are reduced by values from 84% to greater than 99% at 6.9 m BGS, and from 79% to greater than 99% at the deeper depth. These large reductions in concentration are consistent with the high rates of soil-phase mass removal observed for these levels. Note that these calculations do not include DEC or UND, which were not detected in the aqueous samples.

The concentration reductions at the 5.7 m (BGS) depth are not as pronounced as those observed for the other depths, which is consistent with the higher post-remediation soil-phase concentrations at the shallower depths (11). This result is also consistent with the operation history of the CSF and related events. The water table was maintained at about 5.5 m BGS during the CSF, but water-level fluctuations of up to 0.5 m occurred during the experiment. Thus, soil within the sampling influence of the 5.7m-deep level of the MLS array may not have been in contact with the flushing solution during the entire CSF, limiting the extent of remediation. Additionally, fluctuations of the water table during water-flushing and the tracer tests, which were conducted in the period between the CSF and the post-remediation static groundwater sampling episode, may have resulted in contamination of the upper portion of the treatment zone by immiscible liquid from the unremediated zone.

The average of all samples from all three levels (including those from the 6.3m level) show an increase in the concentrations for 3 of the 8 targets (Table 2). However, this is due to the effect of including the 5.7m-level concentration values. That is, since these concentrations were generally larger than those at the deeper depths, and did not show a concentration reduction because of the reasons listed above, they bias the overall average.

Table 1   Aqueous Concentration and Cyclodextrin-Induced Solubility Enhancements in Extraction-Well Effluent for Target Contaminants

Compound	Final Water-flush Concs. (mg/L)[a]	Initial CSF Concs. (mg/L)	Solubility Enhancement[c]
UND	0.0005[b]	14.6	25500[e]
DEC	0.0002[b]	1.2	6890[e]
TMB	0.0046	4.2	989
m,p-XYL	0.0083	7.1	899
DCB	0.325	160.3	493
EB	0.0086	2.6	309
TCE	0.1850	47.6	264
o-XYL	0.0144	3.3	236
NAP	0.0715	10.9	148
TOL	0.2437	22.6	93
BENZ	0.2002	d	d
TCA	1.0006	d	d

a. Average for last 3 days of water flush.
b. Estimated from listed solubility and calculated mole fractions as explained in (11).
c. Enhancement determined from initial CSF conc./final water-flush conc. in E52 only.
d. Not reported due to analytical difficulties (as described in text).
e. Conservative estimate based on calculated mole fraction and solubility (11).

Table 2. **Average Static Ground-Water Contaminant Concentration (mg/L)**

Target	5.67 m (BGS)[a]	6.89 m (BGS)[b]	8.11 m (BGS)[c]	All Samples[d]	5.67 m (BGS)[e]	6.89 m (BGS)[f]	8.11 m (BGS)[g]	All Samples[h]
	Pre-	Flushing			Post-	Flusing		
TCA	4.175	1.677	0.168	1.887	3.183	0.089	0.007	0.653
Toluene	3.247	1.440	1.011	1.824	3.229	0.002	0.006	0.678
DCB	2.174	0.808	0.230	1.023	6.551	0.077	0.048	1.366
m,p-XYL	1.388	0.679	0.368	0.775	0.388	0.002	0.001	0.084
TCE	1.892	0.244	0.201	0.729	0.002	0.005	0.002	0.003
NAP	0.200	0.071	0.073	0.115	0.255	0.011	0.012	0.063
EB	0.064	0.028	0.107	0.065	0.175	0.001	0.001	0.034
o-XYL	0.082	0.048	0.013	0.045	0.235	0.006	0.001	0.055
Benzene	0.061	0.008	0.022	0.028	0.365	0.007	0.001	0.073
TMB	0.028	0.017	0.012	0.018	0.144	0.001	0.001	0.033

a. 11 samples   e. 7 samples
b. 12 samples   f. 11 samples
c. 12 samples   g. 11 samples
d. 37 samples   h. 37 samples

TCA (1,1,1-trichloroethane), DCB (1,2-dichlorobenzene), XYL (xylene), TCE (trichloroethane),
NAP (naphthalene), EB (ethylbenzene), TMB (a,2,4-trimethylbenzene)
BGS = below ground surface
DEC & UND were below analytical detection limits in aqueous samples and are not reported.

CONCLUSIONS

Based on the results presented above, it is clear that the CSF technology was successful in enhancing the remediation of the immiscible-liquid contaminated site. As an enhanced-solubilization technology, the CSF technology is not as aggressive as mobilization-based flushing technologies. However, there are many situations wherein mobilization may not be a viable option. In such cases, there are several attributes of cyclodextrin that may offer advantages compared to other agents for solubilization-based enhanced flushing.

Cyclodextrin is a glucose(sugar)-based molecule and, therefore, is essentially non-toxic. Cyclodextrin is generally nonreactive with soils, which means it can be used efficiently and can be recovered from the subsurface relatively easily. In addition, the solubilization power of cyclodextrin is insensitive to pH and ionic strength. Furthermore, ongoing research indicates that cyclodextrin can enhance the biodegradation of organic contaminants. The use of remediation technologies based on enhanced solubilization will rarely be economically feasible unless the solubilization agent is recovered and reused. Cyclodextrin can be separated and recovered relatively easily from groundwater effluent typical to petroleum and solvent contaminated sites.

In total, the CSF technology using cyclodextrin appears to be a viable candidate for use in source-zone remediation, and merits continued development, testing, and evaluation. Each solubilization agent (e.g., surfactant, cosolvent, cyclodextrin) has advantages and disadvantages, and it is unlikely that any one agent will be the most appropriate for all circumstances. Therefore, selecting which solubilization agent to use for a specific site should be based on a site-specific analysis of all applicable factors.

ACKNOWLEDGEMENTS

This material is based upon work funded by the U.S. Environmental Protection Agency under Cooperative Agreement No. CR-822024 with funds provided through the Strategic Environ. Research and Development Program (SERDP) of the U.S. Department of Defense. This document has not been subjected to Agency review; mention of trade names or commercial products does not constitute endorsement or recommendation for use. We would especially like to thank George Reed and Cerestar USA Inc. for donating the cyclodextrin used in the field experiment. We thank Reid Leland, Mark Bastache, and Dan Colson, who provided valuable support as OU1 site managers. In addition, we thank several University of Arizona graduate students for helping with the field experiment.

REFERENCES

1. National Research Council, **Alternatives for Ground Water Cleanup**, National Academy Press, Wash. D.C., 1994.
2. Bedient, P.B.; Holder, A.W.; Enfield, C.G.; Wood, A.L. In: **Field Testing of Innovative Subsurface Remediation and Characterization Technologies** (this volume).
3. Bender, M.L. and Komiyama, M., **Cyclodextrin Chemistry**, Springer-Verlag, New York, NY, 1978.

4.   Wang, J-M.; Miller, R.M.; Brusseau, M.L. In: Proc. of the American Chemical Society National Meetings, Environmental Chemistry Div., Anaheim, CA, April 2-7, 1995, Vol. 35, American Chem. Soc., Washington, D.C.

5.   Brusseau, M.L., Wang, X., and Hu, Q. Environ. Sci. Technol. 1994, 28, 952-956.

6.   Hu, Q.; Brusseau, M.L. Water Resour. Res. 1995, 31, 1637-1646.

7.   Brusseau, M.L.; Cain, R.B.; Nelson, N.T.; Hu, Q. In: **Field Testing of Innovative Subsurface Remediation and Characterization Technologies** (this volume).

8.   Wang, X.; Brusseau, M.L. Environ. Sci. Technol. 1995, 29, 2632-2635.

9.   Wang, X.; Brusseau, M.L. Environ. Sci. Technol. 1993, 27, 2821-2825.

10.   Bizzigotti, G.O.; Reynolds, D.; Kueper, B.H. Environ. Sci. Technol. 1997, 31, 472-478.

11.   McCray, J.; Brusseau, M.L. Environ. Sci. Technol. 1998 (in press).

12.   Cain, B.C.; Brusseau, M.L. In: Proc. of the American Chemical Society National Meetings, Environmental Chemistry Div., San Francisco, CA, April 13-17, 1997, Vol.37, American Chem. Soc., Washington, D.C.

13.   McCray, J.M.; Bryan, K.; Johnson, G.R.; Brusseau, M.L. In: **Field Testing of Innovative Subsurface Remediation and Characterization Technologies** (this volume).

14.   Beadle, J.B., J. Chromat. 1969, 42, 201-212.

15.   Kondo, H., Nakatani, H., Hiromi, K., Carbohyd. Res. 1976, 52, 1-29.

Chapter 10

Field Test of Cyclodextrin for Enhanced In-Situ Flushing of Multiple-Component Immiscible Organic Liquid Contamination: Comparison to Water Flushing

John E. McCray[1], Kenneth D. Bryan[1], R. Brent Cain[1], Gwynn R. Johnson[2], William J. Blanford[1], and Mark L. Brusseau[1,2]

[1]Departments of Hydrology and Water Resources and [2]Soil, Water, and Environmental Science, University of Arizona, Tucson, AZ 85721

A pilot-scale field experiment was conducted to compare the remediation effectiveness of an enhanced-solubilization technique to that of water flushing for removal of multicomponent nonaqueous-phase organic liquid (NAPL) contaminants from a phreatic aquifer. This innovative remediation technique uses cyclodextrin, a sugar (glucose)-based molecule, to enhance the apparent aqueous solubility of organic contaminants. The cyclodextrin solution significantly increased not only the apparent solubility for several target contaminants, but also the rate of dissolution. As a result of these effects, the time required for cleanup of NAPL contamination at this field site may be greatly reduced by using cyclodextrin-enhanced flushing. For example, it was estimated that more than 70,000 pore volumes of water flushing would be required to remove the undecane mass that was removed in the 8-pore volume cyclodextrin flush, and for trichloroethene, which exhibited the smallest solubility enhancement, about 350 pore volumes of water flushing would be required.

Pump and treat is currently the conventional method for remediation of subsurface contamination. However, traditional pump-and-treat techniques are considered to be ineffective for removal of nonaqueous phase liquids (NAPLs) from the saturated zone (*1,2,3*). The aqueous solubilities for many organic compounds are often very small, typically in the order of milligrams per liter. Therefore, large amounts of water must be flushed through the subsurface to remove the contaminant mass, even under conditions of equilibrium mass transfer between the NAPL and aqueous phases.

Contaminant concentrations in extracted water usually decrease rapidly after initiation of pumping and then level off asymptotically to values that are typically well below the equilibrium concentrations (*3*). Thus, removal of NAPL-phase contamination by dissolution into the aqueous phase occurs much more slowly than is expected because of these apparent mass-transfer limitations. Accordingly, the presence of residual NAPL can serve as a long-term source of ground water contamination, and may severely limit the attainment of remediation goals at many sites (e.g., *1, 3*).

Due to the well-documented limitations of pump-and-treat, alternative methods for remediation of NAPL-contaminated sites are a focus of current research (*1*). Enhanced in-situ flushing is an innovative remediation technique that is currently attracting a great deal of attention. Solubility-enhancement agents increase the apparent solubility of organic pollutants in the flushing fluid, thereby increasing the mass removed per pore volume flushed. Cosolvents (such as alcohols) and surfactants are examples of reagents that have been proposed for this purpose (*2*). Numerous laboratory studies have been conducted to investigate the effectiveness of surfactants and cosolvents (see *4, 5* for summaries). However, porous media heterogeneity, NAPL distribution, and NAPL composition are likely to be much more complicated at the field scale than in a carefully controlled laboratory study. For this reason, many researchers agree that pilot-scale field experiments are needed to gain the information necessary to design and conduct successful full-scale remediations (*6,7,8*). However, few field tests of enhanced-solubility, in-situ flushing technologies have been conducted under well-defined and controlled conditions.

The purpose of this paper is to compare the effectiveness of cyclodextrin flushing (i.e., enhanced pump and treat) to that of water flushing (i.e., basic pump and treat) for remediation of NAPL contamination under essentially identical hydrodynamic and hydrogeologic conditions. Cyclodextrin is an oligosaccharide (sugar-based) molecule that complexes organic contaminants, thereby enhancing their apparent solubility.

The cyclodextrin-flushing and water-flushing experiments were conducted between June and August 1996, at a Comprehensive Environmental Response, Compensation, and Liability Act (CERCLA) site located at Hill Air Force Base (AFB) Utah. This study is one of several treatability studies of innovative remediation technologies that have been conducted at Hill AFB under the Strategic Environmental Research and Development Program (see Bedient et al. (*9*)). This experiment is the first to use cyclodextrin, a complexing sugar, as an alternative flushing agent for subsurface remediation at the field scale. For an overview of using cyclodextrin as an enhanced-solubility agent for subsurface remediation, as well as for initial results on the mass removal effectiveness of the cyclodextrin flush, see McCray and Brusseau (*10*), and Brusseau et al. (*11*).

Site Characterization

Hydrogeology. The aquifer at the site consists of fine-to-coarse sand interbedded with gravel and clay stringers, and has an average thickness of about 8-9 m. The effective porosity within the test cell is approximately 20% (*10,12*). The saturated portion of the aquifer has a horizontal hydraulic conductivity of about 2.5 x 10^{-2} cm/sec based on conservative-tracer test data (*12*). Underlying the sand-gravel unit is a relatively impermeable clay unit. The average vertical hydraulic conductivity of the clay unit is less than 10^{-7} cm/sec based on constant-head permeability testing of core samples collected from the unit (*13*).

Contaminant Location and Distribution. The treatment cell used for the remediation experiment was emplaced within a former chemical disposal pit in what is considered to be a source area of water-immiscible organic contaminants (free phase or residual NAPL present). The NAPL mixture is comprised mainly of petroleum hydrocarbons and spent solvents, and is considered to be less dense than water. The NAPL appears to exist as residual saturation within most of the treatment area, and is smeared throughout the saturated, as well as unsaturated, portions of the aquifer as a result of water table fluctuations. The initial NAPL saturation was estimated as approximately 12.6% based on results of a partitioning tracer study (*12*).

Of the many compounds in the NAPL, 12 "target contaminants" were chosen prior to the experiment for the purpose of evaluating remediation effectiveness. These targets are: trichloroethene (TCE), 1,1,1-trichloroethane (TCA), naphthalene (NAP), o-xylene (o-XYL), m,p-xylene (p-XYL) measured as p-xylene, toluene (TOL), benzene (BENZ), ethylbenzene (EB), 1,2-dichlorobenzene (DCB), 1,2,4-trimethylbenzene (TMB), decane (DEC), and undecane (UND). These compounds were selected to provide a representative subset of the NAPL constituents of concern present within the study area. We estimate that these compounds comprise slightly less than 10% of the total NAPL within the cell (*14*). The remainder appears to consist primarily of higher molecular weight jet-fuel components, other solvents, and perhaps relatively insoluble, pitch-like components. A more detailed description of the field site hydrogeology and contaminant distribution may be found in Bedient et al. (*9*).

Methods

Complexing Sugar Flush (CSF) Experiment. The enhanced-remediation technology tested in this study is termed a "Complexing Sugar Flush" (CSF), where cyclodextrin is the complexing sugar. Cyclodextrin is a glucose-based compound that is similar to household corn starch. These lampshade-shaped molecules are unique in that they have a hydrophobic, non-polar interior and a hydrophilic, polar exterior. Relatively non-polar organic contaminants partition to the interior of the molecule (i.e., form inclusion complexes), while the highly polar exterior of cyclodextrin provides the molecule with a large aqueous solubility (approximately 50% by mass). These properties allow cyclodextrin to significantly increase the apparent aqueous-phase concentrations of organic contaminants.

The cyclodextrin derivative used for this experiment was hydroxypropyl-β-cyclodextrin (HPCD), which does not significantly reduce the interfacial tension of the NAPL-water interface and, therefore, is unlikely to mobilize NAPL (*14,15*). Thus, NAPL removal resulting from cyclodextrin-flushing occurs via an enhanced-solubilization process, and not by mobilization. While mobilization can enhance removal of NAPL from the subsurface, it can be difficult to capture all mobilized

NAPL during remediation. Mobilized DNAPL, for example, could migrate to an uncontaminated region deeper in the aquifer (e.g., *8,16,17*). Remediation techniques based on mobilization, therefore, may not be appropriate under certain circumstances.

The cyclodextrin solution was pumped into the contaminated aquifer using a horizontal flow field. The effluent was collected and subjected to selected treatment and disposal operations. An enclosed cell was used for the experiment to minimize migration of normally sparingly-soluble contaminants that could experience enhanced solubilization and transport in the presence of cyclodextrin, and to facilitate mass-balance and performance assessment. The 3-m by 5-m area cell was enclosed by sealed 9.5 mm-thick sheet-pile walls (*18*), which were driven about 2 m into the low-permeability clay layer. The top of the clay was approximately 8-9 m below ground surface (BGS) in the treatment cell.

A schematic of the cell layout is shown in Figure 1. A line of four injection wells and a line of three extraction wells were used to generate a steady-state flow field that was normal to the lines of wells. The injection and extraction wells (5.1 cm diameter) were fully screened over the saturated thickness. Injection and extraction flow were generated with peristaltic pumps (Master Flex I/P with bench-top controller, Tygon LFL tubing from Cole-Parmer) using a separate pump head for each well. Two piezometers inside the cell were used, along with the injection and extraction wells, to monitor water levels during the test.

Technical-grade cyclodextrin (Cerestar USA Inc., Lot 8028) was used for the experiment. This dehydrated material was comprised of approximately 90% HPCD and 10% production byproducts (mainly hydrated ash). The cyclodextrin was then mixed with potable water to achieve a dissolved HPCD concentration of 10.4wt% using an 80,000-L mixing tank with four top-mounted mixers. About 8 pore volumes of the 10wt% cyclodextrin solution (approximately 65500 L total) were pumped through the cell at a rate of approximately 0.8 pore volumes per day, or 4.54 L/min. Tubing degradation in the peristaltic pump system caused flow rate variations of up to 10%; however, variations of greater than 5% rarely occurred for more than two hours. The flow rate was monitored and adjusted to maintain the water table in the enclosed cell at about 5.5 ± 0.3m BGS.

After the eighth day, flow was interrupted for one day to investigate the potential for rate-limited dissolution. Brief results of this flow-interruption experiment will be discussed in a forthcoming section (detailed results may be found in McCray (*14*)). Two more days of cyclodextrin flushing was conducted after the flow-interruption period. At the end of the experiment, about 12 pore volumes of cyclodextrin-free water were flushed through the cell to remove the cyclodextrin. Aqueous samples were collected at each extraction well (E51, E52, E53 on Figure 1) to monitor for target-contaminant and cyclodextrin concentrations. Samples were collected hourly for the first 36 hours of the experiment, and every 3 to 4 hours thereafter.

Core samples, which were collected at various depths throughout the treatment cell before and after the experiment, were analyzed for the total resident concentrations of each target contaminant. The pre-remediation cores were collected while drilling injection and extraction wells, and from other boreholes. The post-remediation cores were collected at locations as close as possible to the pre-

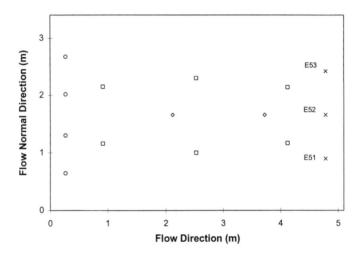

FIGURE 1. Schematic (plan view) of treatment cell, well locations, and soil core sampling locations (O = injection wells, × = extraction wells, ◊ = other pre-remediation soil samples, = post-remediation soil samples).

remediation cores (see Figure 1), within the limitations of the drilling equipment. The cores were collected in sections using a hollow-stem auger with 120 cm-long sample cylinders (USAF, 1995). Only those samples that were collected from the treatment zone were used for this analysis.

Water-Flush Experiment. A water-flush (WF) experiment was performed prior to the cyclodextrin experiment to simulate pump-and-treat remediation for comparison with the CSF. Approximately 19.5 pore volumes of water were flushed through the cell during the water flush at a rate of 0.8 pore volumes per day (4.54 L/min). This pumping period was not continuous; flow was interrupted twice. The first interruption (no-flow) period was imposed after about 17 days of pumping and lasted for 21 days. Pumping was then reinitiated and continued for about 5 days, after which a 4-day no-flow period was imposed. Pumping was then resumed and continued for 3 days, after which the CSF was initiated without interrupting flow.

Aqueous samples were collected from the extraction wells during the experiment and analyzed for concentrations of the target contaminants. It is well known that water-flushing typically results in a rapid decrease in extracted contaminant concentrations as the first pore volume, which may contain relatively high concentrations associated with near-equilibrium conditions, is removed from the subsurface. Concentrations will then typically exhibit (within sample variance) some asymptotic profile wherein concentrations are smaller than the equilibrium values. The concentrations at the end of the water flush (and just prior to the CSF) are thus assumed to be a conservative estimate of the concentrations that would be observed during a long-term pump-and-treat operation.

The CSF was initiated during the asymptotic-concentration period such that steady flow conditions, including a constant water-table elevation, were maintained during the transition between the WF and CSF. Using this procedure, direct comparisons of contaminant mass-removal effectiveness can be made for the two technologies under nearly identical hydrogeological conditions and contaminant distributions. This type of comparison is desirable because it provides direct evidence of the improvement in remediation performance achieved by the CSF compared to pump and treat. It is evident that optimal pump-and-treat remediation design, as well as optimal enhanced-solubility flushing techniques, may not always use the continuous, constant flow-rate design used in this experiment. However, for the purposes of comparison, both remediation technologies must have a common basis.

Analytical Methods. All aqueous-phase extraction-well samples were collected with no head space in 40-mL glass vials with teflon-lined caps. Five-mL portions of the samples were then transferred to 20-mL glass head-space vials (Teflon-lined septum and crimp cap) and analyzed by gas chromatography with a flame ionization detector (GC-FID) (Shimadzu, GC-17A) fitted with a capillary column and a head-space autosampler (Tekmar, 7000). When storage was necessary, samples were stored in the dark at 4°C for periods up to several months. Results from several analyses indicate that volatilization and biodegradation (or other means of mass loss) were not significant during the storage period. With the method described above, quality assurance-quality control standards were met for 10 of the 12 compounds for the CSF samples.

Benzene exhibited soil-phase concentrations several orders of magnitude less than the other target compounds, and presented analytical difficulties for samples containing cyclodextrin due to a coeluting compound. Similar analytical difficulties related to coelution also occurred for TCA in samples containing cyclodextrin. Thus, CSF results for BENZ and TCA are not reported in this research. The compounds BENZ, DEC and UND could not be consistently detected in water-flush samples, and are thus not reported for the water-flush analysis.

During coring, soil samples were placed into vials containing methylene chloride (extractant) and acid, and delivered to the Environmental Laboratory at Michigan Technological University for analysis. Samples were stored at 4 °C until analysis. Prior to analysis, the samples were allowed to equilibrate to room temperature and then were sonicated for 15 minutes. After sonication, the samples were centrifuged at 2300 revolutions per minutes for 15 minutes to separate the phases. A 1.0 ml aliquot of the methylene chloride phase was added to a 1.8 mL auto-sampler vial with a Teflon-lined septum and crimp cap. Then, 2 µL portions of the samples were analyzed by a gas chromatograph (GC) (Hewlett Packard 5890 with capillary column)-mass spectrometer (Hewlett Packard 5970), equipped with an automatic sampler (Hewlett Packard ALS 7673). Using this method, quality-assurance-and-control (QA-QC) standards were met for all 12 compounds.

Results And Discussion

Contaminant Elution. As expected, the concentrations for most contaminants decrease by a factor of about 10 to more than 100 within the first few days of the water flush (e.g., see Figure 2), and maintain relatively asymptotic levels thereafter. Slight increases in concentration appear to have occurred after each flow interruption period (this will be discussed in a forthcoming section). The amount of pumping required to reach asymptotic concentration levels varies among the contaminants, but is generally one to four pore volumes. Thus, the equivalent of between 15-18 pore volumes were flushed during the asymptotic-concentration period.

The initial contaminant concentrations are relatively high, probably because the system was static for several weeks prior to the experiment, which provided sufficient time to achieve equilibrium between the NAPL and water phases. The subsequent small, asymptotic concentrations are typical for pump-and-treat operations (2,3) and, as will be discussed in a forthcoming section, probably result from rate-limited dissolution between the NAPL and aqueous phases. The mass removed during the WF was relatively small for the targets; thus, concentration reductions due to decreasing mole fractions are not of significance. Decreasing concentrations may also result if the flushing fluid, due to porous-media heterogeneities or NAPL relative-permeability effects, experiences limited contact with the NAPL (bypass flow). However, as will be discussed later in this chapter, bypass flow is not thought to be the primary constraint on dissolution for this experiment. Generally, an enhanced flushing technique would not be used until asymptotic-concentration levels are reached with conventional water flushing because pump-and-treat remediation may be nearly as

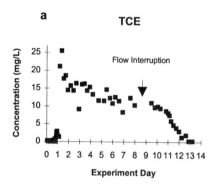

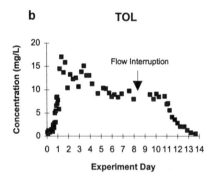

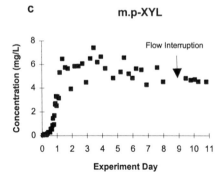

FIGURE 2. Flux-averaged extraction-well concentrations for selected contaminants during the water flush. The horizontal axis represents the number of days for which pumping was conducted. Flow interruption 1 consisted of a 21-day no-flow period, flow interruption 2 consisted of a 4-day no-flow period.

effective as (and less costly than) enhanced-solubility remediation schemes in the early stages of pumping.

The elution curves for selected contaminants obtained during the CSF are presented in Figure 3. The concentrations are flux-averaged values obtained from the concentrations of all three extraction wells. As will be discussed in a forthcoming section, cyclodextrin induced very large increases in the aqueous concentrations of all the target contaminants, ranging from about 100 to more than 10,000 times the concentrations achieved in the water flush conducted immediately prior to the CSF. The effluent concentrations experience a large, initial increase followed by a decrease to a somewhat constant value for most contaminants. The initial increase occurs simultaneously with the increase in HPCD concentration to 10%. The subsequent decrease in concentrations for most contaminants are believed to be due, in part, to the effect of a decreasing mole fraction (*10,14*)

Remediation Performance. As a direct measure of the effectiveness of the CSF, the concentrations and mass removed of the target contaminants during the CSF are compared to the average concentrations and mass removed during the water flush conducted before the CSF. This comparison allows for analysis of the relative magnitudes and rates of mass removal for the two remediation methods. As stated previously, steady flow conditions, including a constant water table elevation, were maintained during the transition from the water flush to the CSF. Therefore, any concentration enhancement in the extraction well effluent during the CSF was due solely to the effects of the cyclodextrin on the apparent solubility of the contaminants.

The amount of contaminant mass removed from the subsurface during water- and sugar-flushing can be calculated for each contaminant by integrating under the elution curves (Figures 2 and 3). The total amount of mass of each target contaminant removed from the treatment cell during flushing, as well as the percentages of total mass removed by the water flush and by the CSF, are reported in Table 1. The water flush removed significant amounts of mass for some of the more soluble components (e.g., the ones shown in Figure 2). However, most of the mass removal for the water flush occurred during the first several pore volumes. As is apparent from the results in the table, the CSF removed significant amounts of mass for these compounds after the mass removal became negligible during the water flush.

The CSF-induced solubility enhancements for the target contaminants are shown in Table 2. The aqueous concentrations for all compounds listed in the table except TCE were below GC detection limits in the effluents of one or more extraction wells, typically E53 or E51, for the latter part of the water flush. In these cases, the lesser of the GC-method detection limit or the calculated equilibrium concentrations (solubility times calculated mole fraction) was used as a conservative estimate of the maximum water-flush concentrations. Thus, the actual concentration enhancements for these contaminants may have been greater than those reported here due to an overestimation of the water-flush concentrations in the effluents of one or more wells. Mass removal may be inferred from the solubility enhancement. That is, for constant water-flush and CSF concentrations, and when equal volumes are flushed, the ratio of mass removed for the two flushing techniques should approximately equal the solubility enhancement.

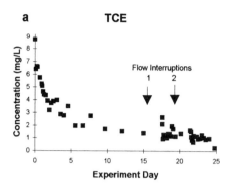

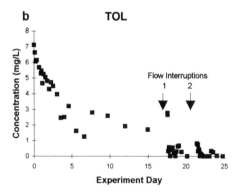

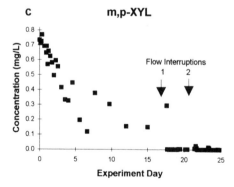

FIGURE 3. Flux-averaged extraction-well concentrations for selected contaminants during the CSF.

TABLE 1. Relative Amounts of Mass Removed by Water Flush (WF)
and Complexing Sugar Flush (CSF)

Compound	Total Mass Removed WF + CSF (g)	% Removed by Water Flush: (First 11 PV)[a]	% Removed by Water Flush: (Last 8 PV)[a]	% Removed by CSF (8 PV)[a]
TOL	994	31.3	0.8	67.9
TCE	919	19.2	1.1	79.8
o-Xyl	131	19.0	< 0.1	81.0
m,p-Xyl	369	8.8	0.1	91.1
EB	103	7.3	0.1	92.6
DCB	5273	4.2	0.1	95.7
NAP	385	2.3	0.2	97.5
TMB	193	2.8	< 0.1	97.6
DEC	54	< 0.01	< 0.01	>99.9
UND	288	< 0.01	< 0.01	>99.9

PV = pore volumes
a. Calculated from total mass removed via extraction wells from aqueous concentration data

The apparent solubilities for the target compounds are greatly enhanced by the cyclodextrin solution. The initial solubility enhancements range from 85 for TCE, to greater than 12,000 for UND. Because solubility enhancements for many of the compounds are minimum values, trends in the enhancements based on contaminant properties are not easily discerned. However, it is apparent from the data in Table 2 that the more-hydrophobic compounds experience a larger solubility enhancement. These results are consistent with those obtained in laboratory studies (14,19). The absolute amount of mass removed for a contaminant depends on the contaminant's aqueous concentration in the extracted cyclodextrin solution (which is a function of the solubility enhancement, the water solubility of the contaminant, and the NAPL-phase mole fraction of the contaminant) and the number of pore volumes flushed (10). For the target contaminants in this study, larger mass removal was generally achieved for the less-hydrophobic compounds due to their higher initial CSF concentrations in the extraction-well effluent and to the relatively short duration of pumping (10).

The contaminant concentrations in the CSF flushing fluid, and thus, the concentration enhancements, generally decline with time and achieve a relatively constant level. Based on partitioning-tracer test analysis (12), a significant portion of the mass (about 44%) was removed prior to reaching the final CSF concentrations shown in Figure 3. The final concentrations are indicative of the long-term mass removal of the remaining NAPL that would be achieved by the CSF. The long-term water-flushing concentrations that would be expected for the conditions of this experiment are the asymptotic values (Figure 2) listed in Table 2. Once effluent concentrations reach an asymptotic level during water flushing, one would expect these levels to remain constant, or to slightly decrease, for a very long time. The long-term mass-removal improvement achieved by the CSF, therefore, can be estimated by comparing the final CSF concentrations to the final water-flush concentrations. This ratio is expressed as the final enhancement value in Table 2. For the target compounds, the advantage of using cyclodextrin as a flushing-solution agent is apparent.

TABLE 2. Aqueous Concentrations and Concentration Enhancements
for Water Flush (WF) and Complexing Sugar Flush (CSF)

Compound	WF Final Conc. [a] (mg/L)	CSF Initial Conc. [b] (mg/L)	[c] Initial Solubility Enhancement	[d] CSF Final Conc. (mg/L)	[e] Final Solubility Enhancement
UND	0.0005[e]	6.0	>12000	5.0	>10,000
DEC	0.0002[e]	1.3	>6500	1.3	>6500
m,p-Xyl	0.0029[f]	6.5	>2241	4.6	>1586
TMB	0.0017[f]	3.7	>2128	2.8	>1610
DCB	0.1162[f]	144	>1240	48	>413
EB	0.0033[f]	2.4	>735	0.8	>245
o-Xyl	0.0067[f]	2.4	>359	1.2	>180
NAP	0.0292[f]	9.2	>315	4.5	>154
TOL	0.0828[f]	17	>205	8.7	>105
TCE	0.2346	20	85	9.1	39

a. Flux-averaged concentration of all 3 extraction wells (during last 3 days of water flush)
b. Initial maximum extraction-well concentration (flux-averaged concentration for all 3 extraction wells) after [HPCD] reached 10%.
c. Solubility enhancement = avg. extraction-well concentration for CSF / avg. extraction-well concentration for WF.
d. Average extraction-well concentration during last day of CSF.
e. Below analytical detection limits in all extraction wells. The aqueous conc. was approximated using the mole fractions times aqueous solubility as a conservative estimate of the steady-state water flush concentration.
f. Below analytical detection limits in one or two extraction wells. The aqueous conc. for these wells was approximated as in e. The actual concentrations for the remaining well(s) are used.

As the NAPL saturations become very small, it is possible that the CSF concentrations may decrease below the final concentrations shown in Figure 3. Nonetheless, compared to water flushing, the amount of mass removed during the 8-pore volume (10-day) CSF alone is substantial (Table 1). For example, assuming that the water-flush concentrations for UND would be maintained indefinitely, more than 70,000 PV (240 years) of water-flushing would be required to remove the amount of UND mass removed during the 8-pore volume CSF. Similar calculations for DCB indicate that more than 5300 PV (18 years) of water flushing would be required after the asymptotic concentration levels are reached. The improvement is significant even for TCE, which has the smallest enhancement, for which about 350 PV (more than one year) of water flushing would be required to remove the mass that was removed in ten days by the CSF. These results clearly indicate that the time required for NAPL cleanup at this field site may be greatly reduced by cyclodextrin-enhanced flushing.

Flow Interruption Experiments. The potential for rate-limited dissolution during the CSF was investigated by interrupting flow for one day after 8 days of flushing. Two flow interruption periods were induced during the water flush (as described previously). During flow periods, less time is available to achieve chemical equilibrium between the NAPL and the flowing aqueous phase than for when flow is stopped. If mass transfer between the NAPL and aqueous phases is rate-limited rather than instantaneous, then an increase in the effluent concentration should be observed once flow is restarted. Figure 3 illustrates that there is not a significant increase in the effluent concentrations of the selected target contaminants after restarting flow (on day 9). Generally, these results indicate that the NAPL dissolution in the presence of the cyclodextrin solution was near equilibrium during the latter stages of the CSF. However, the duration of the no-flow period may not have been sufficient to detect changes in local mass-transfer processes. Thus, these results should not be interpreted to mean that NAPL dissolution was instantaneous everywhere within the cell.

During the water flush there appears to have been a slight increase in concentrations for the most of the target contaminants following the no-flow periods (at about 17 and 22 days in Figure 2, for which the no-flow periods are omitted from the time axis). This result indicates that non-equilibrium dissolution conditions may have existed during the water flush, which was hypothesized earlier. The durations of these no-flow periods were considerably longer than the one for the CSF; but do not appear to have been of sufficient duration to attain the equilibrium concentration values exhibited at the start of the water flush. Another method for determining whether the global NAPL-aqueous phase mass-transfer processes were at equilibrium is described in the following section.

Analysis of Equilibrium Dissolution Behavior. Ideal equilibrium dissolution of NAPL constituents into water may be represented by Raoult's Law:

$$C^{W*}_i = X^N_i S^W_i \tag{1}$$

where C^{W*}_i is the equilibrium aqueous concentration of NAPL constituent i in water, X^N_i is the mole fraction of i in the NAPL phase, and S^W_i is the single-component aqueous solubility of component i in water. For ideal NAPL dissolution into a solution containing an enhanced-solubility agent, a simple modification of Raoult's Law may be used (*14*):

$$C^{A^*}_i = X^N_i \, E_i \, S^w_i \qquad (2)$$

where $C^{A^*}_i$ is the equilibrium aqueous concentration of NAPL constituent i in a solution containing cyclodextrin (the superscript A denotes the aqueous phase), the parameter E_i is the solubility enhancement induced by the cyclodextrin solution (which may be determined in the laboratory from batch solubility experiments). The magnitude of E_i depends linearly on the cyclodextrin concentration and the partition coefficient of the cyclodextrin-organic molecule complex (*19*).

To determine whether the global NAPL-aqueous phase mass-transfer processes within the treatment cell were at equilibrium for various constituents during the water flush and CSF, the aqueous concentrations predicted by equation 1 for water, and by equation 2 for the cyclodextrin solution, can be compared to the measured concentrations. This comparison requires an estimate of the NAPL-phase mole fractions. These mole fractions may be calculated from: measured values for soil-core concentrations, NAPL saturations (*10, 12*), porosity, and soil bulk density (2 g/cm^3); an estimated value for the average molecular weight of the bulk NAPL (180 g/mole); and tabulated molecular properties of the target contaminants. For details of the mole-fraction calculation, the reader is referred to McCray and Brusseau (*10*).

The use of soil-core concentrations, which represent the total contaminant mass in all phases (NAPL, water, and soil), to calculate NAPL-phase mole fractions requires the assumption that all of the contaminant mass resides in the NAPL phase. Total contaminant mass in the cell before and after remediation was calculated from the soil-core concentrations, cell volume, and estimated bulk density (2 g/cm^3). Total aqueous-phase contaminant mass present in the cell before and after remediation was calculated from static-groundwater-sample concentrations (*10,11*). Sorbed (soil-phase) mass was estimated from the static groundwater concentrations and literature values for sorption coefficients (*14*). For this study, it was estimated that more than 98.5% of the mass of each target contaminant resided in the NAPL phase.

Using the calculated mole fractions, theoretical equilibrium aqueous concentrations are calculated for the start of the water flush, the end of the water flush, and the initial phase of the CSF (Table 3). The analysis indicates that the predicted concentrations are generally within a factor of about three to four of the measured ones at the start of the water flush and within a factor of three of the measured concentrations during the initial phase of the CSF. Several researchers (*20,21*) have used a factor of two as the arbitrary standard for ideal, equilibrium dissolution behavior. Given the complexity of this field experiment, a factor of three between measured and expected concentrations would support the assumption of equilibrium dissolution. During the latter phase of the water flush, however, the measured concentrations are significantly below the theoretical predictions based on Raoult's Law. Because the mole fractions at the beginning and end of the water flush are nearly identical, the equilibrium concentrations should also be nearly identical. Thus, this difference is attributed to the effect of rate-limited dissolution during the latter stages of the water flush, as discussed below.

TABLE 3. Measured (C_m) vs. Theoretical (C_{ideal}) Aqueous Concentrations for Water Flush (WF) and Complexing Sugar Flush (CSF)

			Initial WF	End WF	Initial CSF
Compound	E_i	Solubility[d]	C_m/C_{ideal}	C_m/C_{ideal}	C_m/C_{ideal}
TCE	6.5^a	1100	2.3	0.14	2.4
DCB	27^a	137	0.8	0.05	2.0
NAP	30^a	110	1.5	0.20	2.1
TMB	17^a	57	1.5	0.02	2.9
mp-XYL	$24^{b,c}$	170	3.7	0.02	1.6
o-XYL	16^a	152	1.0	0.02	0.4
EB	25^a	150	1.7	0.04	1.1
TOL	10^a	550	1.7	0.03	0.7
TCA	5.8^b	4500	3.6	0.68	g
BENZ	3.5^b	1790	f	f	g
DEC	8650^a	0.02	f	f	0.9
UND	6300^a	0.015^e	f	f	1.8

a. Measured for this work
b. Obtained from K_{ow} vs. K_{cw} correlation (10), updated using recent measurements (14).
c. Average properties for m- and p-xylene assumed for solubility.
d. Representative solubility from values listed in (22,23).
e. Values for UND are average of values for DEC and Dodecane from (23).
f. Not reported; below analytical detection limits.
g. Not reported due to analytical difficulties for samples containing cyclodextrin.

Dissolution Rate Differences Between the Water Flush and CSF and Implications for Remediation. The hydrodynamic conditions and NAPL distributions at the end of the water flush were essentially identical to those at the start of the CSF. Thus, given that dissolution was rate limited at the end of the water flush and near equilibrium during the initial stage of the CSF, it appears that the cyclodextrin solution caused an increase in the rate of NAPL dissolution. This phenomenon may also be illustrated by comparing the solubility enhancements measured in the field (Table 2) to the solubility enhancements measured with batch equilibrium studies (Table 3). The values measured for the batch experiments reflect equilibrium concentrations.

However, under the field conditions, dissolution is an equilibrium process during the CSF, and is rate limited during the water flush. Thus, the field-measured solubility enhancements are generally an order of magnitude larger than those measured from the batch experiments. These results are strong evidence the CSF caused an increase in the dissolution rates for the target contaminants.

Bypass flow (described earlier) may often limit NAPL-water contact during pump-and-treat remediation, and thus result in sub-equilibrium aqueous concentrations. However, if the effects of bypass flow had been significant, then it would be expected that aqueous concentrations measured during the CSF would also have been below equilibrium values. Thus, bypass flow is not thought to have caused significant dissolution limitations during the water flush or the CSF. The cyclodextrin flushing solution did not cause a measurable change in the hydrodynamic conditions at the site, and mobilized NAPL (droplets or emulsions) were not present in extraction-well samples (*10, 14*). Thus, the apparent increase in the dissolution rate is attributed to the impact of the cyclodextrin on the rate of mass transfer between the NAPL and aqueous phases. The mechanisms potentially responsible for this are currently under investigation.

Conclusions

These pilot-scale field experiments were intended to mimic one-pore-volume-per-day remediation scenarios while maintaining strict hydrodynamic conditions that may not be possible in some large-scale remediation efforts. For the conditions under which these experiments were conducted, the CSF is significantly more efficient than water flushing for removing NAPL contaminants from the subsurface. The CSF appears to increase not only the magnitude of the apparent solubility for the target contaminants, but also the rate of dissolution. Thus, cyclodextrin flushing may be useful at sites where dissolution of NAPL into a flushing solution is significantly inhibited due to mass-transfer rate limitations.

Acknowledgments

This material is based upon work funded by the U.S. Environmental Protection Agency under cooperative agreement no. CR-822024 to The University of Arizona with funds provided through the Strategic Environmental Research and Development Program (SERDP). This document has not been subject to agency review; mention of trade names or commercial products does not constitute endorsement or recommendation for use. The authors would especially like to thank George Reed and Cerestar USA Inc. for donating the cyclodextrin used in the field experiment. Reid Leland, Mark Bastache, and Dan Colson deserve thanks for providing valuable support as site managers. Finally, gratitude is expressed to Carl Enfield and Lynn Wood of the EPA Kerr Laboratory for the considerable support they provided for this project.

152

References

1. National Research Council (NRC) (U.S.),. *Alternatives for Ground Water Cleanup*, National Academy Press, Washington, D.C. 1994.
2. Palmer, C.D., and Fish, W. *Chemical Enhancements to Pump and Treat Remediation*, USEPA, Washington D.C., 1992, EPA/540/S-92/001.
3. MacKay, D.M.; Roberts, P.V.; Cherry, J.A. *Environ. Sci. Technol.* 1985, 19(5), 384-392.
4. Shiau, B-J.; Sabatini, D.A.; Harwell, J.H.. *Ground Water*, 1994, 32 (4), 561-569.
5. Augustijn, D.C.M.; Jessup, R.E.; Rao, S.C.; Wood, L. *J. Environ. Engin.*, 1994, 120, 42-56.
6. West, C.C. In *Surfactant_Enhanced Subsurface Remediation: Emerging Technologies*; Sabatini, D.; Knox, R., Harwell, J., Eds.; ACS Sympos. Ser. 594, American Chemical Society, Washington D.C., 1995, 280-285.
7. Gierke, J.S.; Powers, S.E.. Water Environ. Res., 1997, 69 (2), 196-205.
8. Fountain, J.C. Water Environ. Res., 1997, 69(2), 188-195.
9. Bedient, P.B.; Holder, A.W.; Enfield, C.G.; Wood, A.L.. In: *Field Testing of Innovate Subsurface Remediation and Characterization Technologies*; (This Volume), 1998.
10. McCray, J.E.; Brusseau, M.L. *Environ. Sci. Technol.* 1998, (in press).
11. Brusseau M.L.; McCray, J.E.; Johnson, G.R.; Wang, X.J.; Enfield, C., Wood, A.L., In: *Field Testing of Innovate Subsurface Remediation and Characterization Technologies*; (This Volume) 1998.
12. Cain, R.B.; Brusseau, M.L. .*EOS Trans. Amer. Geophys. Union*, 78(46), 1997, F223.
13. Air Force Treatability Study Work Plan, Operable Unit One, 1995, Hill AFB, UT.
14. McCray, J.E. Ph.D. Dissertation, The University of Arizona, Tucson, AZ, 1998
15. Bizzigotti, G.O.; Reynolds, D.; Kueper, B.H. *Environ. Sci. Technol.* 1997, 31(2): 472-478
16. Sabatini, D.; Knox, R.; Harwell, J.H. *Environmental Research Brief,* U.S EPA\600\S-96\002, Washington D.C., 1996.
17. Farley, K.J., Boyd, G.R., Patwardhan, S.,. *Hydrocarbon Contamination in Groundwaters: Removal by Alcohol Flooding*, S. Carolina Water Resour. Institute, Clemson Univ., SC, 1992.
18. Starr, R.C., Cherry, J.A., and Vales, E.S., In: Ontario Ministry of the Environment Technology Transfer Conference, Nov 25-26, Toronto, Ontario, 1991.
19. Wang, X.; Brusseau, M. L. *Environ. Sci. and Technol.* 1993, 27(12), 2821-2825.
20. Lee, L.S.; Rao, P.S.C.;Okuda, I. *Environ. Sci. Technol.* 1992, 26(11), 2110-2115.
21. Cline, P.V.; Delfino, J.J.; and Rao, P.S.C. *Environ. Sci. Technol.* 1991, 25, 914-920.
22. Mongtomery, J.; Welkom, L. *Ground Water Chemicals Desk Reference*, Lewis Pub., Chelsea MI. 1989
23. Mongtomery, J. *Ground Water Chemicals Desk Reference Vol2*, Lewis Pub., Chelsea MI., 1991.

Chapter 11

Field Test of Air Sparging Coupled with Soil Vapor Extraction

John S. Gierke[1], Christopher L. Wojick[2], and Neil J. Hutzler[2,3]

[1]Departments of Geological Engineering and Sciences, [2]Civil and Environmental Engineering, and [3]College of Engineering, Michigan Technological University, Houghton, MI 49931-1295

A controlled field study was designed and conducted to assess the performance of air sparging for remediation of petroleum fuel and solvent contamination in a shallow (3-m deep) groundwater aquifer. Sparging was performed in an isolation test cell (5 m by 3 m by 8-m deep). A soil vapor extraction (SVE) system was installed within the cell perimeter for collecting the sparge vapors. The distribution of twelve target compounds, representing the many volatile and semivolatile chemicals making up the contaminant mixture inside the cell, was obtained from soil and groundwater samples taken before and after sparging treatment. Gas samples were collected during sparging from the SVE offgas and were analyzed for specific target compounds, total hydrocarbons, oxygen, and carbon dioxide. Offgas monitoring and soil sample analyses yielded different, and sometimes contradictory, estimates of sparging performance.

Air sparging is a technique used to remove volatile pollutants from contaminated aquifers (1,2). Basically, the process consists of an air injection system usually within or below the contamination zone (Figure 1). The lateral distribution of air channels depends on the injection rate and pressure, depth of the injection, capillary pressures and permeabilities of the aquifer materials, location of heterogeneities (due to variations in soil properties and the presence of nonaqueous phase liquid contaminants), and degree of anisotropy (3-5). Volatile pollutants will partition into the air channels created by sparging and are carried by advection and diffusion/dispersion into the overlying unsaturated zone (6). Contaminant volatilization rates in a sparge zone have not yet been quantified for field systems. Volatilization rates are dependent upon the distribution of the air channels in the sparge zone, the distribution of the contaminants relative to the air channels, and the rates of dissolution, aqueous diffusion, desorption, and gas/water and gas/nonaqueous-phase mass transfer.

A soil vapor extraction (SVE) system, which induces subsurface gas flow via blower-induced vacuums applied to vents installed in the unsaturated zone, is often coupled with air sparging to capture the vapors emanating from the sparge zone (Figure 1). Soil vapor extraction is a well-established technique with some appropriate design guidance (7). Air sparging is commonly used, but design guidance has been limited because of problems associated with monitoring treatment performance (2,8).

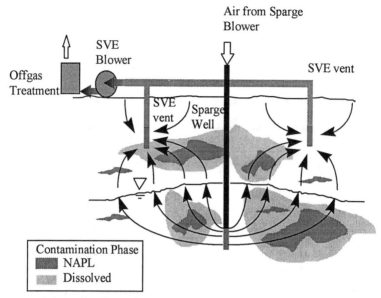

Figure 1. Schematic of a typical air sparging system configuration coupled with a soil vapor extraction system.

Most monitoring efforts have focused on establishing the treatment zone in terms of the distribution of air flow around sparge wells (2).

The extent of the sparge zone, as defined by the distribution of the sparge air channels, is only one design consideration. It is also important to realize that volatilization rates within the sparge zone may vary because of the heterogeneous nature of sparge air flow. Remediation may also occur outside the region where air is flowing as dissolved contaminants can diffuse and may be advected towards (or away from) the sparge zone and oxygen in the sparge air will dissolve into the groundwater and diffuse and advect away from the sparge zone. This study focused on attempting to monitor contaminant removal rates and sparging effectiveness in a field setting.

Quantitative information for ascertaining the contaminant removal efficiency of an air sparging system in a field setting has not yet been reported in the peer-reviewed literature (7). Most studies have either been limited to a technology demonstration or have focussed on observing only the extent to which sparge air emanates from a sparge well (2, 6). Studies of chemical removal associated with sparging that have appeared in the literature, primarily in conference proceedings and government reports, only consider the offgas concentration of the SVE system (Figure 1) or only monitor groundwater concentrations in the treatment zone. Groundwater concentrations are not sufficient to evaluate performance when contamination exists, as it usually does, in sorbed and NAPL phases. Another focus of this study was to ascertain the effect of sparging on contaminant concentrations in an aquifer where a nonaqueous-phase liquid (NAPL) mixture was present.

Description of the Field Site

The treatment performance of air sparging was evaluated in a controlled field test in Chemical Disposal Pit 2 (CDP2) of Operable Unit 1 (OU1) at Hill Air Force Base

(HAFB), Utah, along side performance assessments of 7 other innovative technologies (9). Like the other technology assessments, the air sparging assessment was conducted in an isolation cell (Cell 1), which was located inside the former chemical disposal pit. The pit was used by HAFB until 1974. Although a wide variety of chemicals were dumped at the site, contamination consisted primarily of jet fuel and solvents used for aircraft maintenance at the base (9). The contamination was present mostly as a NAPL mixture that was less dense than water, as the NAPL contained primarily fuel constituents. Since the NAPL was lighter than water, it resided in a smear zone created by natural water table fluctuations and in some locations it floated on the water table. The water table beneath CDP2 fluctuates seasonally because it is a shallow aquifer perched above a low-permeability clay layer. The saturated zone within Cell 1 was isolated from the natural water table fluctuations. Unlike some of the other test cells, floating product was not observed in water wells in Cell 1.

The test cell location was prescribed arbitrarily by the sponsoring agency. Many of the most abundant contaminants in Cell 1 (see **Soil Coring** subsections below) were compounds with vapor pressures less than 1 mm Hg. Remediation by means of volatilization was, therefore, impractical for most of the contamination (1). This test was performed in a site where the conditions with respect to contaminant volatility were challenging for air sparging. After the characterization phase, we learned that the composition of the contaminants in some of the other test cells were richer with higher volatility compounds (C. Enfield, U.S. Environmental Protection Agency, personal communication, 1998). Air sparging likely would have been more effective in those areas.

Field Experiment

A research group from Michigan Technological University conducted the air sparging/soil vapor extraction (AS/SVE) field test at this site in one of eight test cells. The test cells were approximately 3-m wide by 5-m long by 8-m deep and isolated from the surrounding soil by driving interlocking sheet pile (Figure 2) to a depth of 3-m below a confining clay layer. The clay layer was present at an average depth of 8-m below ground surface (bgs). After installation, all sheet pile joints were grouted to hydraulically isolate the cell. A flexible membrane liner was installed across the top surface of the ground within the AS/SVE test cell to prevent short-circuiting of surface air into the SVE system and to enhance the capture of the sparge vapors.

The AS/SVE system utilized 6 SVE vents and 2 sparge wells (Figure 2). The SVE vents were located near the cell walls, fashioned to surround the centrally located sparge wells. During operation of the AS/SVE system, atmospheric air was injected below the water table through the sparge wells, while soil gas was drawn through the SVE vents screened within the vadose zone and directed to a common manifold before treatment by granular activated carbon (Figure 1). Measurements of contaminant removal were made from gas samples collected at this common manifold (Figure 2). Measurements of flow rates for the injection and extraction of air to the cell were made using direct-reading flow gauges. The locations of these gauges provided data on air flow to each sparge well and gas flow from each of two sets of SVE vents (three vents in each set) located along the sides of the cell. Measurements of flow at individual SVE vents were made using an averaging pitot tube and differential pressure gauge. The flow to individual sparge wells and inlet and extraction vents could be controlled using ball valves located in the sparging and SVE piping.

The performance assessment consisted of pre- and post-treatment characterization activities and monitoring during treatment. The pre-/post-treatment characterization was consistent with the other demonstrations and included: soil coring with subsamples analyzed for target compounds, groundwater sampling and analysis of

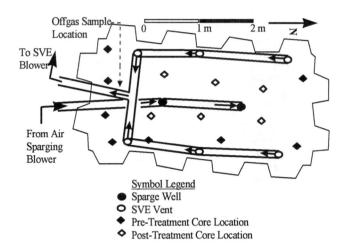

Symbol Legend
● Sparge Well
○ SVE Vent
◆ Pre-Treatment Core Location
◇ Post-Treatment Core Location

Figure 2. Plan view of the AS/SVE configuration and the soil core locations in Test Cell 1 of Operable Unit 1 at Hill AFB, Utah.

target compounds, and partitioning tracer tests. During the treatment phase, soil gas and SVE offgas were monitored for target compounds, total volatile organic chemicals (TVOC) as trichloroethene (TCE), oxygen, and carbon dioxide.

Contamination was present in both the vadose and saturated zones. Because the sparge air, which is injected below the water table, must pass through the vadose zone before removal by the soil vapor extraction system (Figure 1), contaminants present in the SVE offgas may originate from either the vadose or saturated zones. To evaluate air sparging for remediating the saturated zone, the SVE system was operated initially without sparging in an attempt to remove as much of the contamination from the vadose zone as practical. Upon completion of this initial cleanup of the vadose zone, the combined AS/SVE system was operated. Ideally, it would have been preferable to operate the SVE phase until nondetectable concentrations were observed in the offgas for all the constituents. Because of the project time schedule, it became necessary to switch to AS/SVE before complete cleanup of the unsaturated zone. It is not appropriate to extrapolate the observations of the SVE phase in an attempt to quantify the relative contributions of contaminants in the saturated and unsaturated zones because the vapor flow patterns are probably different during the SVE and AS/SVE phases. Therefore, the reported removals for AS/SVE herein are maximum achievable rates as the contribution of continued cleanup of the unsaturated zone during AS/SVE was not accounted for.

Operation of the SVE system commenced on 28 August 1996 and final shutdown of the cell cleanup was on 22 October 1996. The SVE system was operated initially at a flow rate (all gas flow rates are reported herein at standard temperature and pressure, STP) of 6.8 m^3/hr (4.0 standard cubic feet per minute, SCFM) and the flow rate incrementally increased to 34 m^3/hr (20 SCFM) before the start of combined AS/SVE operation. Sparging was started on 20 September 1996, initially at a flow rate of 14 m^3/hr (8.0 SCFM) and incrementally increased up to 37 m^3/hr (22 SCFM) before final shutdown. During the coupled AS/SVE operation, the extraction flow rate was concomitantly increased along with the sparging rate. The extraction rate was adjusted to be approximately 25%, ±7% for one standard deviation, higher (17-46 m³/hr) than the sparging rate to prevent the escape of sparging vapors from the test cell.

Figure 3 depicts the flow history of the sparging system, including two shutdown periods prior to the final stoppage, as a function of the cumulative volume of gas extracted from the test cell. The cumulative volume extracted is used here as a common "time" axis in all graphs. In total, 43,700 m³ of gas (at STP) were flushed through the test cell. Sparging was started after a cumulative volume extracted equal to 13,800 m³, before which treatment of the unsaturated zone was being accomplished by SVE alone. The pore volume in the unsaturated zone was flushed 5600 times (1800 during SVE alone). The pore volume in the saturated zone was flushed 3900 times.

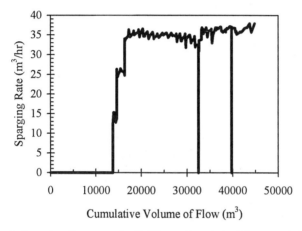

Figure 3. Sparging flow rates (at STP) as a function of the cumulative volume (at STP) of gas extracted from the test cell.

Soil Coring. Soil cores were taken at eight designated locations within the test cell prior to the technology demonstration and at six locations after completion of the demonstration (Figure 2). Cores were collected in consecutive lengths of 1.2 m (4-ft) across the entire depth of the treatment zone (2.7- to 7.6-m bgs). Subsamples of the cores (~5 g) were taken at approximate intervals of 0.6 m (2 ft) and transferred to 40-ml VOA vials containing a water/dichloromethane (DCM) extractant solution. A 1.8-ml aliquot of the extractant was analyzed by gas chromatography/mass selective detector (GC/MSD) for twelve target compounds (Table I) according to a method common to the other performance assessments at HAFB (9).

Using the sampling locations for the cores (Figure 2) and corresponding concentration data, total mass estimates were developed for each constituent for both pre- and post-treatment. The groundwater table (GWT) depth was 4.7-m bgs during the air sparging demonstration, thus mass estimates for the analytes were made for the vadose and saturated zones separately using the GWT as the boundary.

Table I. Vapor Pressure and Aqueous Solubility of Constituent Compounds[a].

Compound	Vapor Pressure (mm Hg)	Saturation Vapor Concentration[b] (mg/L)	Aqueous Solubility (mg/L)
cis-1,2-dichloroethene (1,2-DCE)[c]	<200[d]	<1000	3350
1,1,1-trichloroethane (1,1,1-TCA)	68	510	2170
benzene	50	220	1510
trichloroethene (TCE)	39	290	856
toluene	14	73	435
ethylbenzene	4.4	26	129
m-xylene	3.8	23	125
o-xylene	3.0	18	172
1,3,5-trimethylbenzene (1,3,5-TMB)	1.2	8.1	25.9
1,2-dichlorobenzene (1,2-DCB)	0.55	4.5	17.7
decane	0.52	4.2	5.48
undecane	0.14	1.3	0.719
naphthalene	0.10	0.72	22.9

[a]Values reported by (10) for 12 °C, unless otherwise noted.
[b]Calculated from the vapor pressure.
[c]Was analyzed for only in gas samples but not in soil or groundwater samples.
[d]From (11) for 25 °C.

Offgas Analyses. The SVE offgas was sampled at the combined-well manifold (Figure 2) through a 3-mm diameter, stainless-steel tubing. The sample tubing was directly connected to a Hewlett Packard 5890 (HP 5890) gas chromatograph (GC). The sample tubing was heat traced and held at a temperature of 150°C to minimize adsorption of the organic compounds. A vacuum pump installed in the sample vent of the GC was used to draw the gas sample from the SVE manifold and through gas-sample valves. The GC was configured to measure either TVOC concentrations or the concentrations of the target compounds (Table I).

Since the primary focus of this chapter is to describe the observed chemical removal by volatilization, the methods for gas analyses of contaminants are described below. The methods for analyzing oxygen, carbon dioxide, groundwater and soil samples are of secondary importance, so only the references are given for those methods.

The GC was customized with several features to facilitate measurement of the TVOC and target compound concentrations. Two columns were installed in the gas chromatograph: (1) a HP-624 (60-m long, 0.32-mm diameter, 1.8-μm film thickness) capillary column was used for analysis of the 13 constituent compounds in the offgas; and (2) a HP Retention Gap (uncoated, deactivated, 0.32-mm diameter) capillary was used for the TVOC analysis. A flame ionization detector (FID) was found to have sufficient sensitivity for all the analyses. The GC was equipped with a 10-port sample valve, which was used to apply atmospheric pressure to both ends of the sample loop immediately prior to injection of the gas sample onto the column. This ensured a consistent sample loop pressure, and thereby gas sample volume, independent of the type of method of sample introduction (e.g., Tedlar bag versus sample line connected to the SVE manifold). Oven temperature for the TVOC analysis was isothermal at 103 °C, while for analysis of the target compounds the temperature program was: 35 °C held for 5 minutes, then ramped at 3 °C/min to 80 °C and held for 30 minutes, then ramped at 3 °C/min to 140 °C and then ramped at 10 °C/min to 200°C and held for 5 minutes. Total time for the constituents analysis was 81 minutes.

Standards for the TVOC and target compounds were prepared onsite and immediately prior to use. Trichloroethene was used for calibration of the TVOC analyses. The TVOC standards were prepared in 10-liter or 0.5-liter Tedlar bags at concentrations of 2, 10, 100, 400, 2000, and 4000 ppm_v. A high-concentration stock gas (67,000 ppm_v) was first prepared in a 0.5-liter Tedlar bag by injecting a quantity of liquid TCE, measured in a syringe, into the bag containing a measured volume of nitrogen gas. The TVOC standards were then prepared in nitrogen gas by dilution of the stock gas. A gastight syringe was used to measure and transfer the stock TCE gas into the Tedlar bags used to contain the standards.

Standards containing the 13 target compounds were prepared in 10-liter Tedlar bags at concentrations of 0.5, 5, 50, and 100 ppm_v. The standards were made by injecting a quantity of a liquid stock solution, measured in a syringe, directly into a Tedlar bag containing a measured volume of nitrogen gas. The liquid stock had been prepared previously at a known concentration by dilution into methanol. The volume of gas contained in the Tedlar bags was determined by injecting nitrogen gas at a known flow rate for a measured period of time.

Initially, the gas standards were analyzed daily to check reproducibility, instrument drift, and degradation of the standards. Significant degradation of the TCE and constituent compound standards was observed to occur within 24 hours after preparation. Thus a practice was adopted of analyzing a set of standards immediately after preparation and using this calibration for the next three or four days until another set of standards was prepared. During the days between preparation of standards, the most recently prepared 10-ppm_v TCE standard was analyzed several times daily to monitor for deterioration of instrument performance. The 10-ppm_v standard was chosen for this check since the low-concentration standards showed the least degradation over time. A similar practice was followed for the constituent compound standards except the 5-ppm_v standard was used as the check, and the analysis was performed only once each day.

The SVE offgas was also sampled at the combined SVE manifold with a syringe, which was used to put the sample in a 0.5-L Tedlar bag. The samples in the bags were analyzed for O_2, CO_2, and TVOC concentrations. The TVOC analysis was performed on some samples as confirmation for the samples obtained through the heat-traced tube. For the TVOC analysis, the sample was delivered to the HP 5890 using the vacuum pump as described above. Oxygen and CO_2 were analyzed with a different GC and the data are presented elsewhere (*12,13*). Concentrations reported on a mass per volume basis were initially determined on a volume per volume basis and then converted to standard temperature and pressure (STP) values.

Results

Soil Coring. A summary of the concentrations of the target compounds in the soil samples below the groundwater table taken before and after the AS/SVE treatment is given in Table II. The compounds are listed in order of decreasing vapor pressure. The more volatile constituents (1,1,1-TCA, benzene, TCE, toluene) were not observed in very high concentrations. The low concentrations could have been due to volatilization during sampling, however, groundwater and gas sample analyses confirm that the highest volatility targets (vapor pressure > 10 mm Hg) were not present in measurable quantities. 1,2-DCE was an exception in that it was observed in the offgas during SVE, but it was not a target compound in the soil and groundwater analyses.

The mass of each target constituent in the test cell was calculated by apportioning the concentrations horizontally using the polygonal method (*14*). Depths were subdivided using incremental intervals matching the average sampling depths. The bulk soil density was estimated to be 2600 kg/m^3 based on a gravimetrically measured particle density (2610 kg/m^3) and a cell porosity determined by measuring the changes in water table elevation corresponding to measured volumes of water withdrawals (*13*). Decane and undecane were the most prevalent of the target

Table II. Concentrations and Masses of Target Compounds Determined from Analysis of Soil Samples.

	Pre-Treatment[a]			Post-Treatment[b]			
	Concentrations		Mass[c]	Concentrations		Mass[d]	Mass
Compound	Median	Maximum		Median	Maximum		Change[e]
	(mg/kg)	(mg/kg)	(g)	(mg/kg)	(mg/kg)	(g)	(%)
1,1,1-TCA	0.024	0.35	2	0.010	0.22	>1	n/a
benzene	0.009	0.022	>1	0.012	0.028	>1	n/a
TCE	0.040	0.18	>1	0.019	0.32	1	n/a
toluene	0.017	0.13	4	0.011	0.081	1	n/a
ethylbenzene	0.083	0.96	20	0.005	0.94	9	55
m-xylene	0.12	1.8	36	0.017	1.9	17	53
o-xylene	0.10	4.6	86	0.047	4.4	51	41
1,3,5-TMB	0.85	7.3	180	0.11	12	120	33
1,2-DCB	0.15	2.3	37	0.018	2.0	23	38
decane	1.9	140	2400	0.51	120	1100	54
undecane	22	180	5100	2.1	140	2400	53
naphthalene	0.18	3.4	96	0.015	3.0	39	59

[a]Based on 63 samples from 8 cores.
[b]Based on 48 samples from 6 cores.
[c]Estimate of constituent mass in the saturated zone within the cell based on the 28 soil samples taken from below the groundwater table.
[d]Estimate of constituent mass in the saturated zone within the cell based on the 16 soil samples taken from below the groundwater table.
[e]Percent difference between pre- and post-treatment masses in the saturated zone. Differences denoted as n/a are not reportable as the amounts were too low to quantify.

compounds in the soil and the offgas; 1,2-DCB was also prevalent in the offgas but was not found in as high of concentrations in the soil samples.

The soil cores taken prior to treatment were located near the inside perimeter of the cell (Figure 2). The locations were selected with consideration to utilizing the boreholes for water wells and water sampling devices. There were no significant trends in concentrations horizontally. In general, concentrations increased with depth (*13*). The cores taken after treatment were more centrally located (Figure 2), however, the post-treatment core recoveries tended to be much poorer compared to those taken before treatment. As in the pre-treatment characterization, post-treatment concentrations were higher with depth and there were no significant trends horizontally (*13*).

Offgas Analyses. The SVE offgas was sampled at the combined-well manifold (Figure 2) and analyzed to determine concentrations of TVOC as TCE, and 13 target compounds (Table I). The offgas concentrations of the 13 constituent compounds were measured daily while the TVOC concentrations were measured several times each day. During times of anticipated rapidly changing concentrations (*e.g.*, during a flow change or flow interruption), the TVOC concentrations were measured even more frequently (10- to 20-minute intervals).

Total VOCs as TCE (TVOC). Concentrations measured during operation of the SVE system and combined AS/SVE systems are shown in Figure 4. The initial concentration measurement at the start of operation was 14,500 mg/m^3 (2480 ppm$_v$) as TCE. Although the concentration in the offgas prior to the start of sparging had declined to 430 mg/m^3 (74 ppm$_v$), the rate of contaminant removal from the vadose zone had not yet reached an asymptote.

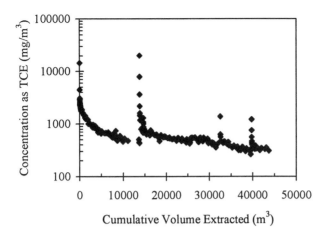

Figure 4. Total VOC concentration in SVE offgas during air sparging.

Upon initiation of sparging a peak concentration value of 20,100 mg/m^3 (3430 ppm$_v$) as TCE was measured. The TVOC content of the offgas declined until upon shutdown the concentration reached 310 mg/m^3 (53 ppm$_v$). The two concentration peaks located at cumulative extracted volumes of 32,500 and 39,800 m^3 both resulted from temporary blower shutdowns (Figure 3). Smaller concentration spikes were observed when sparging rates were increased to 25 and 34 m^3/hr at 14,600 and 16,400 m^3, respectively. The gap in the concentration data just prior to the start of sparging (11,100 to 13,600 m^3 volume extracted) corresponds to when the GC was inoperable due to a broken valve.

Target Compounds. Concentration data for decane, undecane, and 1,2-DCB measured during operation of the SVE system and combined AS/SVE systems are shown in Figure 5. Of the 13 target compounds analyzed from daily gas samples, these three constituents were the only ones present during sparging. Other constituents were present during the SVE phase, but they were not in high enough concentrations to be detected during the AS/SVE phase ([13]). Other chemicals, which were not included with the target list and were not identified, were also present.

Sparging was started at a cumulative extracted volume of 13,800 m^3, which began a gradual increase (over a period of 3 days) in the concentrations of decane, undecane, and 1,2-DCB (Figure 5). The concentration changes of these target compounds did not portray the same abrupt changes observed in the TVOC concentration data (Figure 4). An important difference between the target and TVOC monitoring is the frequency of sampling. Since the TVOC analysis took only a few minutes, it was analyzed for more frequently than the targets, which took 81 minutes for a single run. On average, targets were analyzed only once or twice per day, which was not frequent enough to capture the rapid changes observed by the TVOC monitoring.

The initial gradual increases in target concentrations correspond to incremental increases in the sparging rate (Figure 3), and then reach a maximum 3 days after the maximum sparging rate was achieved. The slow increase in concentrations is much longer than the response of the size of the sparging zone to changes in sparging rate which usually occurs over periods of hours ([15]). The concentration trends are possibly due to a combination of alterations in the sparge zone (*e.g.*, lateral extension of the

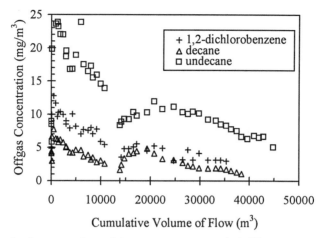

Figure 5. Concentrations of three target compounds (milligrams per standard cubic meter) in SVE offgas collected during SVE and AS/SVE treatment.

sparge air and additional air displacement at higher injection pressures), imperfect contact between the sparge air and the contamination (*i.e.*, diffusion limitations), volatilization of contaminants in the unsaturated zone, and partitioning of contaminants in the sparge vapors into NAPL contamination remaining in the unsaturated zone. It is also possible that the TVOC response is predominantly from substances in the unsaturated zone, hence the more rapid response, or affected by compounds that respond more quickly to changes in flow patterns (*e.g.*, methane production in stagnant gas zones).

It is important to consider, too, that other contaminants beside the targets were being detected by the GC. These were not identified and so they could not be quantified. In particular, three unidentified compounds were observed at levels that were probably quantifiable as their GC-peak area counts were between those of undecane and decane (*13*). Since the contamination was a complex mixture of many compounds, it is also possible that some of the peaks detected by the GC may actually represent more than one compound or even a different compound with the same elution time for the particular GC conditions.

Target Compound Masses Collected in Offgas. The test cell was designed and operated so that all of the volatilized contaminants would be collected in the offgas. Table III summarizes the masses of the three target constituents and TVOC appearing in the offgas during AS/SVE. Because of differences in FID response factors for TCE and the target compounds, the TVOC mass and individual target compound masses can not be compared directly. Nevertheless, it can be concluded that the target compounds were only a small percentage (less than 1-5%) of the total offgas hydrocarbon concentration.

The masses of decane and undecane collected during AS/SVE (Table III) were much less than the calculated masses removed based on the assessment of the soil core analyses (Table II). This discrepancy suggests that other removal mechanisms, such as biodegradation (see next paragraph) or chemical oxidation, were significant or that some of the constituents were removed as a result of aqueous-phase flushing during the partitioning tracer tests (see **Groundwater Analyses** below). Other constituents

Table III. Masses of Constituent Compounds Collected in SVE Offgas

	1,2-DCB (g)	decane (g)	undecane (g)	TVOC (g as TCE)
Mass Removed During SVE Only:	90	53	234	11,000
Total Mass Removed from Cell:	190	123	514	25,000
Mass Removed During AS/SVE[a]:	100	70	280	14,000

[a]Mass Removed During AS/SVE = Total Mass Removed from Cell - Mass Removed During SVE Only.

exhibited significant reductions in soil concentrations, such as o-xylene, 1,3,5-TMB, and naphthalene, but these were not detected in the offgas during AS/SVE. These compounds would have been observed in the offgas if the mass removed was due to volatilization, as their average offgas concentrations would have been above the concentrations of the standards used in the daily calibration of the GC. Moreover, these compounds were present in the offgas during the SVE-only phase (13). However, the mass of contamination estimated in the test cell from soil concentrations is subject to large errors because the contamination was heterogeneously distributed. For example, 1,2-DCB was observed in the offgas in amounts as much as 5 times greater than estimated from the soil core analysis. The observations for decane and undecane should be similar to each other, as both are considered to be aerobically biodegradable and both are constituents of jet fuel. Even though the vapor pressure of 1,2-DCB is only slightly higher than decane and undecane, different removal results and/or mass estimates for 1,2-DCB are likely because: (1) it is not as susceptible to biological degradation and (2) it is not a jet fuel constituent so it was likely to be more heterogeneously distributed.

Monitoring of CO_2 in the offgas showed that soil-gas concentrations of CO_2 were well above atmospheric levels at all times in the vadose zone of the test cell, especially during periods of blower shutdown (12,13). Concomitantly, oxygen concentrations in the soil gas would drop during blower shutdowns. Oxygen consumption rates could be used to estimate a bioconversion rate of the contamination, but this would tend to overestimate the contaminant degradation rate as oxygen is also consumed by the conversion of natural organic matter. Likewise the production rate of CO_2 could be used to estimate the utilization rate of organic compounds, and it tends to provide a lower estimate as CO_2 is a final degradation product and soil and groundwater can act as a sink for CO_2. The average rate of CO_2 production in the test cell was measured to be about 20 moles/day (0.2 moles/m^3-soil/day), which translates, for example, to a stoichiometrically equivalent decane degradation rate for complete mineralization (1 mole of decane yields 10 moles of CO_2) of 280 g/day (1.5 mg/kg-soil/day). Of course, the CO_2 produced was from a combination of other contaminants, both above and below the water table, and soil organic matter as well. Nevertheless, this estimate probably illustrates at least the overall magnitude of the biodegradation potential in the test cell when aerobic conditions exist. The chemical-equivalent degradation rates based on oxygen consumption were consistently higher by a factor of 2 (12).

Groundwater Analyses. Water samples were taken before and after the treatment test, both during and 72-hours after aqueous partitioning tracer tests. The groundwater samples were analyzed by a method of heated-head space with GC/FID, as prescribed in the treatability work plan (9). The aqueous concentrations of all the groundwater samples were between nondetectable (1 ppb) and 41 ppb for all of the targets (13). Spatially and temporally averaged groundwater concentrations were below 4 ppb for all of the targets (Table IV). The aqueous concentrations were so low that there was no statistical difference overall between the pre- and post-treatment samples. Therefore

Table IV. Concentrations of Target Compounds Determined from Analysis of Groundwater Samples and Mass Removal Estimates for Pump-and-Treat (PAT) Based on Concentrations Observed During the Partitioning Tracer Tests.

	Pre-Treatment			Post-Treatment		
	Average[a]		PAT	Average[a]		PAT
Compound[b]	Static[c] (μg/L)	Dynamic[d] (μg/L)	Removal[e] (mg/PV)	Static[f] (μg/L)	Dynamic[g] (μg/L)	Removal[e] (mg/PV)
1,1,1-TCA	2.8	2.8	22	1.3	2.0	15
TCE	0.3	<0.1	<1	3.6	1.3	10
toluene	<0.1	ND[h]	n/a	0.2	0.7	5
m+p-xylene[i]	ND	ND	n/a	1.4	5.6	43
o-xylene	1.1	ND	n/a	1.6	5.5	42
1,3,5-TMB	<0.2	ND	n/a	<0.4	2.7	21
1,2-DCB	2.7	ND	n/a	3.7	6.4	49
decane	ND	ND	n/a	<0.2	ND	n/a
undecane	0.8	ND	n/a	1.1	ND	n/a
naphthalene	ND	ND	n/a	<0.5	1.4	11

[a]Nondetects (ND) were assigned values of zero in the calculation of averages. The detectable concentrations were taken to be above 1 ppb.
[b]1,2-DCE, benzene, and ethylbenzene were not analyzed.
[c]Based on 45 samples taken 72 hours after the pre-treatment PTT.
[d]Based on 37 samples from three extraction wells during the pre-treatment PTT.
[e]Pump-and-treat (PAT) removal rate was estimated by multiplying the average dynamic sample concentrations by the saturated pore volume (1 PV = 7700 L) inside the cell.
[f]Based on 40 samples taken 72 hours after the post-treatment PTT.
[g]Based on 43 samples from three extraction wells during the post-treatment PTT.
[h]ND denotes that all samples were below detection limits, and, hence, a removal rate by PAT was not calculated.
[i]m-xylene and p-xylene were not separated in the groundwater analysis.

the groundwater characterization was not useful in attempting to quantify air sparging performance for the selected target compounds in Cell 1.

The groundwater concentration data was highly variable, and there were no consistent trends during the PTT for the constituents present in detectable concentrations. For many of the groundwater samples the concentrations were very close to the detection limit and are less reliable than both the soil and offgas concentrations.

According to the project work plan (9), the removal rates by AS/SVE for the suite of target compounds were to be compared to a set of baseline removal rates for a conventional pump-and-treat (PAT) system. The groundwater concentrations that were observed during the aqueous partitioning tracer tests were used to estimate removal by PAT on a mass removed per pore volume displaced basis (Table IV). The PAT removals were obtained while flushing the cell with water at a rate of approximately 0.84 pore volumes (7700 L) per day (4.5 liters per minute (lpm)). The removal rate of 1,2-DCB, for example, during AS/SVE was 26 mg per pore volume of sparge air. This rate is greater than what was observed during the pre-treatment PTT as 1,2-DCB was below detection in the extraction-well samples, but less than the rate during the post-treatment PTT, which was 49 mg per pore volume of water. Decane and undecane, on the other hand, were not observed in the groundwater samples during either PTT. This is reasonable given that their aqueous solubilities are the lowest amongst the target compounds (Table I); however, both were observed in the offgas (Figure 5) despite the fact that their volatility is also very low (Table I). The lack of confidence in the groundwater measurements of the target compounds makes the removal rate comparisons to PAT questionable for our tests.

Partitioning Tracer Tests. The partitioning tracer tests (PTTs), which were designed to measure the volume of NAPL contamination in the test cell, yielded conclusive results for only one of the partitioning tracers: 2,2-dimethyl-3-pentanol (*13*). The 2,2-dimethyl-3-pentanol data were judged to be conclusive in that its mass recovery was above 90%, and the other tracers (*n*-hexanol and 6-methyl-2-heptanol) experienced degradation during both the test and sample storage (*16*). Nearly complete recoveries were obtained for the two nonpartitioning tracers, methanol and bromide, which were used to estimate the average pore water velocity. Even though the residence times of the methanol and bromide were different by only 10%, this difference translated into differences of more than 18% in the calculated average NAPL saturations (*i.e.*, ±0.013). Overall, the estimated average NAPL saturation in the saturated zone of Cell 1 was 0.070 (±0.010 for one standard deviation) based on the pre-treatment PTT and 0.079 (±0.011) based on the post-treatment PTT. The precision of the tracer tests was inadequate for ascertaining a significant change in contaminant volume before and after treatment (*16*).

Conclusions

For this air sparging/soil vapor extraction treatability study, offgas chemical concentration data and soil concentrations were the most useful for assessing the performance of air sparging and soil vapor extraction. Contaminant mass removals were observed by monitoring the offgas collected by the soil vapor extraction system and by pre- and post-treatment soil coring analyses. The mass collected of a target constituent in the offgas was typically less than the mass difference between pre- and post-treatment soil-core analyses. The discrepancies suggest two possible occurrences: (1) removal mechanisms other than volatilization were occurring, and (2) the soil concentrations determined from the cores may not accurately represent the actual subsurface mass distributions. For a baseline comparison, contaminant removal rates corresponding to pump-and-treat were determined from monitoring the groundwater extraction wells during the partitioning tracer tests. In general, contaminant removal rates by volatilization due to air sparging were higher than what could be achieved with a conventional pump-and-treat system. The addition of oxygen to the soil and groundwater appeared to enhance contaminant removals, as evidenced by elevated carbon dioxide concentrations in the soil gas. The amount of carbon dioxide collected in the offgas suggests that aerobic biodegradation was significant, and maybe even greater than removal by volatilization for the hydrocarbon constituents. The offgas concentrations of the predominant target analytes (decane, undecane, and 1,2-dichlorobenzene) showed trends in the chemical removal by volatilization as a result of changing the treatment from soil vapor extraction alone to air sparging coupled with soil vapor extraction and due to increases in the sparging rate. Total volatile organic chemical concentrations in the offgas also responded to flow changes, but the duration of the changes were short (less than a day) in comparison to the duration of the changes in target concentrations (a few days). The volume of nonaqueous-phase liquid contamination in the test cell was measured using aqueous-phase partitioning tracer tests before and after treatment. The precision of the tracer tests was not sufficient to measure a reduction in the contaminant volume due to air sparging.

Acknowledgements. This work was funded by the Strategic Environmental Research and Development Program, jointly sponsored by the U.S. Department of Defense, Environmental Protection Agency, and U.S. Department of Energy. The authors gratefully acknowledge Drs. C. Enfield (USEPA, Cincinnati, OH), A. L. Wood

(USEPA, Ada, OK), and J. S. Ginn (Hill Air Force Base, UT) for their assistance with the conduct of this study. In addition, we are grateful to Mr. D. L. Perram (MTU) and Ms. J. M. Muraski-Smith (Montgomery Watson Consultants, Chicago, IL) for their assistance collecting and analyzing some of the data presented herein. The comments and suggestions of the two anonymous reviewers are also greatly appreciated.

References

1. Bausmith, D. S.; Campbell, D. J.; Vidic, R. D. *Water Environ. Technol.* **1996,** *8*, 45.
2. Nyer, E. K.; Suthersan, S. S. *Ground Water Monit. Remediation* **1993,** *13*, 87.
3. McCray, J. E.; Falta, R. W. *Ground Water* **1997,** *35*, 99.
4. McCray, J. E.; Falta, R. W. *J. Contam. Hydrol.* **1996,** *24*, 25.
5. Unger, A. J. A.; Sudicky, E. A.; Forsyth, P. A. *Water Resour. Res.* **1995,** *31*, 1913.
6. Clayton, W. S.; Bass, D. H.; Ram, N. M.; Nelson, C. H. *Remediation* **1996,** *6*, 15.
7. Gierke, J. S.; Powers, S. E. *Water Environ. Res.* **1997,** *69*, 196.
8. Hein, G. L.; Gierke, J. S.; Hutzler, N. J.; Falta, R. W. *Ground Water Monit. Remediation.* **1997,** *17*, 222.
9. Montgomery Watson Consultants, Hill Air Force Base, Utah, Draft Phase II Work Plan for Eight Treatability Studies at Operable Unit 1, April 1996.
10. Daubert, T. E.; Danner, R. P.; Sibul, H. M.; Stebbins, C. C. *Physical and Thermodynamic Properties of Pure Chemicals: Data Compilation* (core with 5 supplements); Taylor and Francis: Bristol, PA, 1995.
11. Verschueren, K. *Handbook of Environmental Data on Organic Chemicals*; Van Norstrand Reinhold Co., Inc.: New York, 1983, 2nd Edition; pp. 1310.
12. Muraski, J. M. In-Situ and Laboratory Measurements of Biodegradation Rate Constants for a Sandy-Gravel Soil Contaminated with Petroleum and Chlorinated Hydrocarbons, M.S. Thesis, Dept. Civil & Environ. Eng., Mich. Technol. Univ., Houghton, MI, 1997.
13. Wojick, C. L. Comparison of Several Alternative and Traditional Methods for the Performance Assessment of an Air Sparging/Soil Vapor Extraction Field Test, Ph.D. Dissertation, Dept. Civil & Environ. Eng., Mich. Technol. Univ., Houghton, 1998.
14. Isaaks, E. H.; Srivastava, R. M. *Applied Geostatistics*; Oxford University Press, Inc.: New York, NY, 1989, pp 238-241.
15. Lundegard, P. D.; LaBrecque, D. *J. Contam. Hydrol.* **1995,** *19*, 1.
16. Sanders, D. L. Laboratory Studies of Aqueous Partitioning Tracer Tests for Measuring NAPL Volumes, M.S. Thesis, Dept. Geol. Eng. & Sci., Mich. Technol. Univ., Houghton, MI, 1997.

Chapter 12

Performance Assessment of In-Well Aeration for the Remediation of an Aquifer Contaminated by a Multicomponent Immiscible Liquid

William J. Blanford[1], E. J. Klingel[2], Gwynn R. Johnson[3], R. Brent Cain[1], Carl Enfield[4], and Mark L. Brusseau[1,3]

[1]Departments of Hydrology and Water Resources and [3]Soil, Water, and Environmental Science, University of Arizona, Tucson, AZ 85721
[2]IEG Technologies Corporation, Charlotte, NC 28269
[4]Andrew W. Briedbach Environmental Research Center, U.S. Environmental Protection Agency, Cincinnati, OH 45268

A pilot-scale test to evaluate the performance of a vertical recirculation well equipped with an in-well air stripper was conducted at Hill AFB, Utah, in an aquifer contaminated with petroleum and chlorinated solvents. During the two months of operation, the air stripping system was found to remove more than 26% of the combined mass of ten representative contaminants from water passing through the well. The cell-wide performance was evaluated by comparing the contaminant concentrations for aquifer core samples collected before and after the test and by comparing the average immiscible liquid saturations determined with partitioning tracer tests conducted before and after operation. The net magnitude of remediation was low (<1%) due to the low aqueous concentrations of the predominant treatable contaminants at the site and the impact of the vertical gradients on immiscible liquid mobilization.

Most sedimentary aquifers are composed of predominantly horizontal layers of differing hydraulic conductivities. Attempts to extract contaminants from such systems using horizontal flushing (e.g., typical pump and treat) are inhibited by preferential flow through the higher permeability layers, which limits the remediation of the lower permeability layers. Thus, removal of contaminants from these layers is often dependent on diffusive mass transfer, which can be significantly rate limited. It is possible to enhance contaminant removal in such systems by inducing flow perpendicular to the strata. This may be accomplished with a vertical circulation well (VCW).

Vertical circulation wells are single wells screened at two or more separate intervals. Water is extracted at one interval and injected into another. In systems without a confining layer between the screens, water flows from one screen to the other along parabolic flow lines. A higher ratio of horizontal to vertical conductivity and greater distance between screens have been shown to extend the flow-lines farther from the well *(1, 2)*. More complicated VCW systems include the addition of multiple wells *(3)*.

Researchers at the University of Oklahoma conducted model simulations, column studies, and a field demonstration of enhanced removal of TCE, PCE, and recalcitrant jet fuel components using a vertical surfactant flush *(4, 5)*. They recovered 95% of the injected surfactant while increasing PCE removal 40 fold and that of jet fuel components 90 fold compared to the horizontal-flow pump-and-treat systems at the site. In conjunction with the vertical circulation well, four other wells were used for monitoring and hydraulic control.

The injection of air into a well can drive circulation through the well, strip volatile contaminants from the water, and increase the dissolved oxygen content of recirculated water, which may enhance the potential for aerobic biodegradation as a secondary means of remediation. Researchers with the USEPA from 1982 to 1985 evaluated several options including airlift pumping with and without in-well diffused aeration (sparging) and electrical submersible in-line pumps coupled with in-well defused aeration for remediation of groundwater contaminated by volatile organic compounds *(6)*. They found that the combination of the submersible pump coupled with in-well aeration was effective in removal of volatile organic compounds and that aeration may provide a useful treatment technique on a short-term emergency basis for vital production wells.

If contaminants are treated within the well, the treated water can be reinjected into the aquifer, thereby creating a vertical recirculation system. Treating contaminated water within the recirculation well offers an alternative to more common techniques involving above-ground treatment of extracted groundwater. One method for separating volatile contaminants from water that is adaptable to recirculation wells is air stripping. The coupling of an air-stripping reactor within a vertical recirculation well has been termed a vacuum vaporizer well (UVB) or in-well aeration (IWA). Less common in-well treatment systems include bioreactors, activated carbon adsorption, and the use of materials such as contact resins and iron filings.

Vertical recirculation systems have been applied to more than 300 sites world wide because of its advantages of in-situ treatment, low operational costs, and ease of use. Herrling et al. *(1, 7)* describe the use of UVB systems to remediate aquifers contaminated by gasoline and chlorinated solvents. Herrling et al.*(8)* discuss the potential uses of vertical circulation wells for free product recovery, soil vapor extraction, bioventing, and soil flushing.

This report presents the results of a test conducted to evaluate the ability of a vertical recirculation well, equipped with an IEG Technologies 150 mm canister in-well aeration system, to remediate a portion of the aquifer at Operable Unit 1 of Hill Air Force base in Utah. The system was operated from July 26 to September 23, 1996. This test was part of a large project designed to evaluate the performance of several innovative remediation technologies *(9)*.

Methods

Operable Unit 1 at Hill AFB has a history of disposal and periodic burning of used solvents and jet fuel in unlined pits. This has contaminated the surficial unconfined aquifer with chlorinated compounds, jet fuel components, and high molecular weight poly nuclear aromatic hydrocarbons (PAHs). Cell 2 is thought not to be located directly beneath the former chemical disposal pits, but within the migratory paths of aqueous and free-phase contaminants.

The aquifer is composed primarily of sand, with gravel and cobbles mixed with clay and silt stringers. It is underlain by a regional clay aquitard at 8 to 8.5 meters (m) below ground surface. A 3 by 5 meter rectangle of sheet piling was emplaced through the aquifer into the clay aquitard. A vertical recirculation well was installed in the center of the test cell accompanied with other wells and multilevel samplers to evaluate performance (see Figure 1). Pre-remediation coring within cell 2 determined that the depth to clay on the south side was 7.8 m below ground surface (bgs), 8.2 m bgs in the middle, and 8.3 m bgs on the north side. The static ground water level was 6.7 to 6.9 m bgs outside the cell and was elevated to 5.8 m bgs inside the cell.

The basic instrumentation of the cell is similar to others at the site *(9)*. Four injection wells at the southern end and three extraction wells at the northern end of the cell were installed for conducting horizontal partitioning tracer experiments common to all test cells. All of these wells are screened from 4.9 to 7.9 meters below ground surface (bgs). The four wells nearest the corners of the cell are 6.4 cm internal diameter and the other three are 5.08 cm. Twelve multi-level sampling systems (MLSs) were installed between the injection and extraction wells. Each MLS has five sampling points spaced every 0.46 meters from 5.8 to 7.6 meters bgs, which provides a three-dimensional sampling array of 60 points.

The IWA well was installed with a hollow stem auger. The auger had an inside diameter of 24 centimeters and an outside diameter of 31 centimeters. The location of the IWA was previously cored and backfilled with sand. Placing this required core at the same location as the IWA minimized the disturbance of the aquifer system and provided a pilot hole for the larger auger used to drill the IWA borehole.

The vertical recirculation well, which contained two screened intervals, has an inner diameter of 15.5 centimeters. The screens of the vertical recirculation well have a 15.2 cm (6 inch) internal diameter and are wire wrapped screens fabricated of galvanized low-carbon steel with round wire (Irragator model, Johnson of Minneapolis, Minnesota). The openings between the wires are 8mm giving 0.18 m^2 opening per meter of screen. The top screen is 0.61 m long, the middle casing is 1.07 m long, and the bottom screen is 0.46 m long, and rests atop a 0.15 m deep sump. The top screen starts at 5.5 m bgs and the bottom screen ends at 7.6 m bgs. The total coverage of saturated aquifer is 2.14 meters. A solid casing made of schedule 40 high density polyethylene connects the top screen to the surface. Two PVC monitoring wells, 2.54 cm inside diameter, were placed within the same borehole as the VCW, one at each screen. Based on experience at the site, 20-40 sand was used for backfill around the screens to enable a productive interface with the aquifer. Micro-bentonite chips were placed surrounding the middle casing between the layers of sand to deter vertical flow within the borehole.

The IWA system employed at OU1 was designed and built by the IEG Technologies Corporation (Charlotte, North Carolina). A schematic of the in-well system is shown in Figure 2. Initial modeling studies were conducted to optimize the size of the IWA and the position of the vertical recirculation well within the test cell. A small pilot-scale IWA system of only 150 mm in diameter was used because of the small size of the test domain (3x5x2 m). This demonstration-scale system may be inherently less efficient in removing contaminants from recirculated water than the

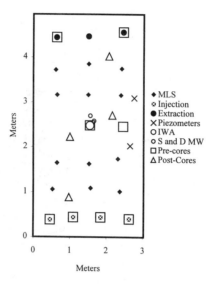

Figure 1 Cell Instrumentation and Coring Locations

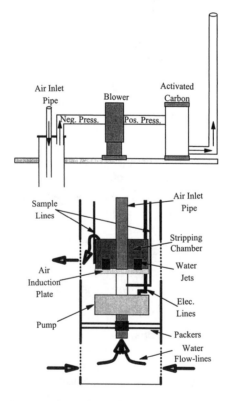

Figure 2 Above Ground and In-well Systems

normal full-sized systems currently used. An in-line pump was used to withdraw water from the lower portion of the aquifer into the lower screen, and up through a pipe within the solid casing separating the two screens. This pipe was surrounded by dual rubber packers to limit recirculation of water within the well. After passage through the pump, the water continued up through three pipes into a stripping chamber. The treated water then spilled over the edge of the stripping chamber, and exited into the top portion of the aquifer through the top screen. This design is a "standard flow" IWA, in which air and water flow concurrently upward within the casing.

A schematic of the above ground system is presented in Figure 2. Air flow through the system is induced via a blower, which withdraws air from the well casing. The resulting negative pressure causes air to flow through the air-stripping chamber from an in-take pipe in the well cap. Air enters through a pinhole plate at the bottom of the cylindrical (45cm deep and 10cm wide) stripping chamber. Water from the pump enters the chamber through three 1 cm diameter jets and mixes with the air. The vacuum created within the well casing by the blower is sufficient to force atmospheric air to overcome the weight of the column of water within the stripping chamber and induce flow through the intake pipe. Air and contaminated water turbulently mix within the stripping chamber, stripping volatile contaminants from the water into the air. The contaminated air is drawn to the surface due to the negative pressure within the well-casing. The contaminated air exits the blower, under pressure greater than atmospheric, and is flushed through a drum of activated carbon for treatment. The contaminant-depleted air is vented to the atmosphere through a 3.8 meter high stack.

Of the many detectable organic compounds at the site, twelve were chosen for use in assessing technology performance. The twelve include common chlorinated solvents such as TCA and TCE, fuel components such as the aromatics, benzene and toluene, and alkanes such as decane and undecane. These compounds have large differences in aqueous solubilities, volatilities, biodegradation potential, and other characteristics (see Table I). The potential remedial performance of the IWA is limited to those compounds with suitable aqueous solubility and volatility. Thus, compounds such as benzene and TCE are potentially more easily removed than decane. The actual removal effected for any specific compound will be further mediated by the complex nature of the physical, chemical, and biological interactions occurring within the aquifer.

Results

Characterization of Groundwater Entering and Exiting the IWA. Thirty pairs of samples were taken from the IWA well to measure the target contaminant concentrations entering and leaving the well during the technology demonstration. Samples were collected from a sampling line located just prior to the air stripper and a sampling line located after the stripper (see Figure 2). The majority of the target concentrations within these samples were near method detection and quantifiable thresholds. For the purposes of determining mean concentrations entering and exiting the IWA, samples with measured target concentrations below detection levels are given the value of the detection level and samples with measured concentrations above the detection level, but below the quantifiable level are assigned the quantifiable level. This is a conservative approach to determining performance because more exiting samples have target concentrations within these ranges. Because the samples were not taken at

Table I. Target Contaminant Results

Target Contaminants	Aqueous Solubility[a] (mg/L)	Henry's Coeffecients[a] (atm-m3/mol)	Theoretical stripping efficiency[b] Air Out/Water In (g/g)	Initial Coring		Final Coring		Div. F./I.	Initial Static Aqueous[d]			Final Static Aqueous[d]		
				Avg.[c] ug/g	St. Dev.[c] ug/g	Avg.[c] ug/g	St. Dev.[c] ug/g		Avg.[c] ug/L Est.	St. Dev.[c] ug/L Est.	St./Equ	Avg.[c] ug/L Est.	St. Dev.[c] ug/L Est.	St./Equ
Benzene	1790	0.00548	98.2%	0.008	0.007	0.010	0.007	125%	1.4	1.3	0.39	0.9	0.4	0.27
TCA	1550[a]	0.016	99.4%	0.019	0.042	0.028	0.023	148%	4.5	3.3	1.00	4.7	6.1	0.92
TCE	1100	0.0099	99.0%	0.044	0.130	0.023	0.016	53%	2.1	1.7	0.27	2.2	1.7	0.70
Toluene	515	0.0067	98.5%	0.142	0.179	0.110	0.096	77%	1.2	0.6	0.07	0.6	0.3	0.06
o-Xylene	152	0.0053	98.1%	1.65	1.43	1.43	1.20	87%	0.6	0.7	0.01	0.3	0.1	0.01
p-Xylene	180	0.007	98.6%	0.681	0.696	0.871	0.615	128%	0.9	1.4	0.04	0.4	0.4	0.02
Ethyl-Benzene	152	0.0066	98.5%	0.242	0.209	0.321	0.217	133%	0.5	0.7	0.07	0.2	0.2	0.03
DCB	137	0.0012	92.1%	0.704	0.582	0.919	0.458	131%	12.6	23.8	0.92	1.4	1.2	0.10
Tri-Methyl-	57	0.0057	98.2%	3.50	2.34	4.22	1.78	120%	0.5	0.8	0.01	0.2	0.2	0.01
Naphthalene	31.7	0.00046	81.8%	1.15	1.19	1.71	1.10	149%	12.0	20.0	1.92	0.7	0.3	0.09
Decane	0.009	0.187[a]	99.9%	42.3	32.0	48.6	15.9	115%						
Undecane				77.5	85.3	99.3	51.6	128%						
Total				127.9		157.5		123%	36.3		0.5	11.7		0.2

[a]Reference (17):Aqueous Solubility and Henry's Coefficients, TCA solubility range 900-4500 at 20C, Decane's Henry's coef. is calculated
[b]Theoretical Equilibrium Volatilized calculated from Henry's Coefficients, Vapor Pressure, and the flux, temperature, and pressure values for Air and Water for the Air Stripping Chamber.
[c]Core and Static Aqueous contaminant concentrations are arithmatic means and standard deviations of all samples.
[d]Initial and Final Static Aqueous Concentrations are from the same locations: 9 samples from 6.9 m bgs and 10 samples from 7.8 m bgs.

equal intervals, a time-weighted mean concentration was computed. TCA is the only target for which a significant fraction of samples had concentrations above detection limits. The results for the other compounds are strongly biased in that only a few of the thirty samples are above the quantifiable threshold. These few highly concentrated samples occurred early and late in operation and contributed more than 75% of the computed mass entering the IWA, but account for less than 10% of the total time of operation.

The percent of each target volatilized was computed by subtracting from unity the result of dividing the exiting mean concentration by the entering mean concentration. Using the mean of the measured air flux, temperature, and pressure conditions, the estimated water flux of 5 L/min, and assuming that air entering the stripping chamber contains no target compounds, a theoretical maximum amount volatilized can be computed (Table I):

$$\text{Stripping Eff.} = \frac{M_{\text{air out}}}{M_{\text{water in}}} = \frac{1}{1 + \dfrac{V_w RT}{V_{\text{air}} K_H}}$$

where V_w is the estimated volumetric water flow rate (m³/hr) passing through the IWA, V_{air} is the averaged measured volumetric air flow rate (m³/hr) passing through the IWA, R is the gas constant (8.21×10^{-5} atm m³ mol⁻¹K⁻¹), T is the average temperature in degrees kelvin, K_H is Henry's constant scaled for air pressure (atm m³ mol⁻¹). Water temperature is assumed to be equal to air temperature due to the large air-to-water volume ratio within the chamber (238 to 1).

The theoretical maximum amount volatilized is greater than the computed values for all the targets except naphthalene, for which theoretical and computed values are nearly the same. TCA, which is considered the most reliable target for measuring stripping efficiency because it has the lowest coefficient of variation of the ten measured targets, has a computed volatilization of 11% compared to a theoretical maximum of 99.4%. This translates into a stripping efficiency of 11.4% for TCA, indicating that air and water do not reach equilibrium within the chamber for TCA. Figure 3 shows the entering and exiting concentration versus time for the study.

An estimate of the time required for the system to remove the target contaminants from the cell is useful to evaluate potential performance. An initial mass of each target can be estimated from initial core concentrations and aquifer properties. The removal of individual contaminants from the cell can be estimated by combining the average influent water concentration, the average stripping efficiency, the estimated water flow rate through the cell, and the total time of operation. If TCA removal continued at the same average rate as occurred during the study, it would take more than two years of operation of the IWA to completely remove TCA from the treatment zone.

Characterization of Air Entering and Exiting the IWA. Air flow was monitored periodically at the air inlet pipe, before and after the blower, and within the stack. Pressure within the air ducts was monitored with pressure gauges while temperature, relative humidity, and flow rate were monitored with the model 452 Instant Action Anemometer (Testo Incorporated, Flanders, New Jersey). Combining the measurements of volumetric flow-rate, temperature, pressure, and relative humidity generates a calculation of water flux for each measurement location. These results indicated the net removal of 17 ml/min of water, or more than 600 liters during the time of operation.

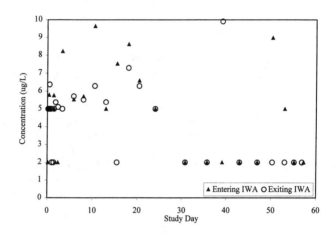

Figure 3 TCA Aqueous Fluxes Through the IWA

The contaminant load in the air was monitored with a portable photo-ionization detector (PID). The PID was calibrated with a reference gas of 100 mg/L iso-butanol and the local atmosphere approximately thirty meters from the site to negate potential background interferences. Results of monitoring of the air stream by the PID are shown in Figure 4. Because the PID is calibrated solely to iso-butanol (MW = 124 grams/mole), the concentration has been combined with flow rate, pressure, and temperature to yield contaminant mass flow measurements as equivalent grams of iso-butanol per day. To estimate overall mass removed, calculations are made as per day for a single measurement by assuming each reading is an average condition for the time period that spans halfway between the previous and subsequent samples. Although humidity and temperature naturally fluctuate with the hour of the day, these influences were relatively small and precautions were taken to limit biases.

Table II shows the total equivalent mass of iso-butanol measured at each sample point along with the average for each sample. The intake concentration rarely rose above zero, with more than 2/3 of the readings taken from the intake being zero. Based on these data, approximately 18 g of equivalent iso-butanol (EIB) is estimated to have entered the cell through the air inlet. The measurements after the blower, which are believed to be more accurate because it is under positive pressure and therefore less subject to dilution, provide a value of 303 grams of EIB mass removed. Subtracting the influent from the effluent air contaminant loads produces a calculated net mass removal of 285 grams in equivalent mass of iso-butanol.

Characterization of Core Samples Before and After IWA Operation. Contaminant concentrations present in cores collected before and after operation were used as one method to evaluate the performance of the IWA system. During coring, soil samples were placed into vials containing methylene chloride (extractant) and acid. They were analyzed at the Environmental Laboratory at Michigan Technological University by a gas chromatograph (GC) mass spectrometer. Table I lists the average and standard deviation of the concentrations for the twelve target compounds for both pre- and post-remediation coring. Initial conditions were determined from eight cores collected with the installation of wells (see Figure 1 for locations). The initial amount of each target contaminant is roughly inversely proportional to its aqueous solubility. For example, the two most insoluble targets, decane and undecane, compose 33% and 60% respectively of the total mass of target compounds. Except for TCE, the target compounds were distributed predominantly from 5.64 to 7.16 meters below ground, with diminishing amounts below this level. Assuming the data obtained from pre-remediation cores accurately reflect the distribution of contaminants within the cell, the total mass of the twelve targets within the cell is calculated to be 6.3 kg.

The post-cores were collected from locations closer to the middle of the cell (see Figure 1). Due to limited recovery from the coring process, only 14 samples exist from the treatment zone. Samples were taken from the four cores at 5.8, 6.4, 7.0, and 7.6 meters bgs depths. Averaging the concentration of each target for each depth profile produces a relatively uniform distribution for most contaminants in the post-cores.

Samples from similar depths are used to compare the pre- and post-coring results. These include samples from 5.8, 6.4, 7.0, and 7.6 meters bgs for the post and samples from 5.6, 5.9, 6.2, 6.6, 7.0, 7.3, and 7.6 meters bgs depths for the pre-remediation coring. Comparisons indicate that the final average concentration is greater

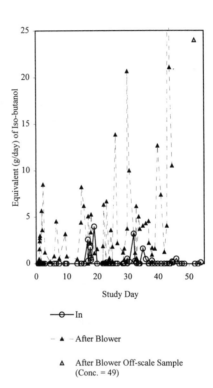

Figure 4 PID Fluxes

than the initial average concentration for nine of the 12 compounds. Post-remediation cores have lower average concentrations for TCE, toluene, and o-xylene.

Table II. IWA Air Stream Parameters

Sampling Location	In	Out of Well	After Blower	Stack
Num. of Samples	75	76	76	72
Avg. Flowrate (L/min)	1189	1864	1318	1227
Avg. Flowrate (Std L/min)	1103	1588	1163	1118
Avg. Temp (°C)	25.9	24.1	43.6	32.4
Avg. Pressure (kPa)	94.2	84.3	95.2	94.2
Avg. Humidity (%RH)	30.4	79.6	27.1	48.8
Avg. Water Flux (ml/min)	32.9	106.3	50.1	48.2
Avg. PID (ppm EIB)[a]	0.24	1.82	3.01	1.75
Total (g EIB)[a]	18	255	303	180
Total (ml EIB)[a]	14	205	244	145

[a]EIB is Equivalent mass as Iso-butanol

The fact that the post-cores appear to be generally more contaminated than the pre-remediation cores indicates a non-uniform distribution of contamination within the cell. Most of the pre-cores were close to the cell perimeter whereas the post-cores were collected from the center of the cell. Given that the center of the cell received the greatest treatment, the best direct comparison would be the two pre- and two post-remediation cores located near the middle of the cell. The two pre and post-cores are located within 0.4 meters of the treatment well. For these two sets of cores, the number of targets experiencing reductions in concentrations increases to five of the twelve, but there is no significant pattern showing greater removal.

During the operation of the IWA, the hydraulics of the vertical recirculation system mobilized and redistributed immiscible liquid. For example, the flow away from the top screen of the IWA moved large quantities of floating immiscible liquid to wells 2252 and 2253, where it was not previously observed. More than 7 liters of the immiscible liquid was collected from the two wells. Accumulation was also observed at other wells within the cell, but in thicknesses less than 2 cm. It is possible that the mobilization and smearing of the immiscible-liquid influenced contaminant distributions during the study, thereby obscuring contaminant removal associated with operation of the IWA system.

Partitioning Tracer Tests. Interwell partitioning tracer studies were performed in cell 2 before and after the operation of the in-well aeration system. Through the chromatographic separation of chemical species, these tracer tests allow both the detection and estimation of an immiscible-liquid saturation (S_n) located within the swept-zone of a porous medium *(11-15)*. The tracers used in the test are either preferentially retained by the immiscible-liquid phase or behave conservatively. Using the travel times and immiscible-liquid/water partition coefficients of these tracers, an effective immiscible-liquid saturation can be calculated for the targeted portion of contaminated porous medium. Comparisons of the preliminary and post estimated immiscible-liquid saturations indicates the amount of immiscible liquid solubilized and removed by the IWA. These studies allow a much larger volume of aquifer to be characterized, compared to other method such as soil cores and groundwater sampling.

As seen in Table III, the results of the cell 2 pre-remediation partitioning tracer test demonstrate a relatively uniform immiscible-liquid distribution across the cell (from wells 2251 to 2253). The S_n values range from 7.4 to 9.7% for the three wells, with a swept-volume weighted averaged of 8.9%. This value translates to 525 L of NAPL and 5366 L of water within the flow field for the pre-remediation study. Table III gives the pore volume, travel times for the conservative tracer (bromide), S_n, and immiscible-liquid volumes for both the pre- and post-remediation partitioning tracer tests.

Table III. Cell 2 NAPL Partitioning Tracer Tests

Extraction Well	Pre-Partitioning Study				Post-Partitioning Study			
	2251	*2252*	*2253*	*Wt. Avg. or Total*	*2251*	*2252*	*2253*	*Wt. Avg. or Total*
Pore Volume (L)	1800	2050	1520	5370	1500	2780	1270	5550
Travel Time (days)	1.24	1.41	1.04	1.25	1.03	1.92	0.87	1.44
Partitioning (S_n%)	9.7	9.3	7.4	8.9	16.7	7.7	14.0	11.6
NAPL Volume (L)	193.4	210.2	121.5	524.6	300.7	231.9	206.7	728.3
Tracer	bromide (370), ethanol (997), pentanol (988), 2,2-dimethyl-3-pentanol (395), hexanol (945)				bromide (307), ethanol (1107), pentanol (522), 2,2-dimethyl-3-pentanol (406), hexanol (991), 6- methyl-2-heptanol (477)			

The post-partitioning tracer test was conducted after the operation of the IWA system. The flow rate and depth to the water-table were essentially identical to those used in the pre-remediation study. Comparisons of the results indicate an increase in S_n from the pre-study to the post-remediation study. This apparent increase may originate from the redistribution of immiscible liquid within the cell as discussed above, and an increase in size of the pore volume observed for the post-partitioning tracer test. Similar apparent increases in immiscible liquid saturation from the pre- to post-partitioning tracer studies were observed for the air sparging test at OU1 *(16)*.

Characterization of Groundwater Prior to and After the IWA Operation. To evaluate the ability of the system to remove specific contaminants, aqueous groundwater samples were collected from the aquifer before and after operation of the IWA. Static conditions were maintained for 72 hours in an attempt to establish equilibrium distributions of the target compound within the cell. To determine the initial groundwater concentrations, 10 samples were drawn from the MLSs at 5.9, 6.9, and 7.8 m bgs just prior to operation of the IWA.

Table I shows the results of the initial static sampling, including the average and standard deviation of the aqueous concentration of the ten most soluble targets. To evaluate if the static aqueous samples are in equilibrium with the immiscible-liquid contamination in the cell, the average static concentration is divided by the expected equilibrium concentration, which is estimated by *(10)*:

$$X_i = \frac{C_i \rho_b V_c MW_n}{S_n \theta \rho V_c MW_i}$$

where X_i is the mole fraction of the component of the NAPL, S_i is the aqueous solubility, C_i is the average core concentration (initial cores are used for comparison with initial statics), ρ_b is the bulk density, which is best approximated at 1.6 g/cm^3, V_c is the cell volume(which cancels out), MW_n is the average molecular weight of the NAPL, which based on the work of McCray and Brusseau *(10)*, is best approximated to be 180 g/mole for this site, S_n is the NAPL saturation, θ is the porosity (0.18), ρ_n is the NAPL density (extracted NAPL from cell 2 had a density of 0.885 g/ml), and MW_i is the molecular weight of the target of concern. The results of these calculations are presented in Table I as the ratio of average static to computed equilibrium concentration.

The ratio of measured to expected shows that the three major constituents in the pre-remediation static samples, naphthalene, DCB and TCA, appear to be close to equilibrium. Conversely, the five alkyl benzenes have very small ratios. This result may not mean, however, that these compounds are actually far from equilibrium. The low concentrations could reflect preferential biodegradation of these more labile compounds. The larger aromatic and the chlorinated target compounds are likely to be less prone to biodegradation than the alkyl benzenes. The intermediate ratios observed for TCE and benzene, which are generally less biodegradable than the alkyl benzenes, may result from volatilization losses during sampling and storage.

The mean aqueous concentrations measured for the 19 post-remediation static samples were lower than those of the initial static samples for eight of the ten detectable target contaminants. The comparison of measured to expected concentration are similar to the initial static data, with the exception of DCB and naphthalene, of which showed apparent nonequilibrium for the final samples. As a result, the distribution of the contaminants changed from the initial samples. In the final static round, the three most soluble, benzene, TCA, and TCE, comprise the majority (67%) of the total dissolved load, whereas DCB and naphthalene did for the initial statics.

Conclusions

The operation of the In-well aeration system at OU1 removed small amounts of contamination from the aquifer. Monitoring of dissolved target contaminants entering and exiting the recirculation well showed a net reduction in the ten contaminants that could be detected. Monitoring of the air stream with a portable photo ionization detector (PID) showed that contaminants in the water entering the well partitioned into the air stream. Combining the readings of the PID, which was calibrated to iso-butanol (124 grams per mole), with temperature, pressure, and flow-rate measurements of the air stream resulted in calculated values of 18 and 303 grams of contaminant mass entering and exiting the IWA, respectfully. Further, the air concentrations did not decrease over the two months of operation, indicating that the system was still removing mass. If the calibration of the PID to iso-butanol accurately reflects the contaminant load within the off gas air stream, then the total mass removal of contaminants is approximately 285 grams. This amount is nearly 24 times more than the amount of the target compounds calculated to have been volatilized by air stripping. However, not including decane and undecane which could not be detected in aqueous samples, the target contaminants are estimated to compose less than 0.1% of the total contaminant mass within the treatment zone. Thus, the higher PID air stream mass removal calculations are not unexpected because many other volatile and semi-volatile compounds were likely removed during air stripping.

To measure the net effect on contamination within the cell, aquifer coring, partitioning tracer-test data, and groundwater samples were collected before and after operation of the IWA. The partitioning tracer tests and core comparisons show increases in contamination present within the treatment zone. The total of the average concentration of the twelve target contaminants increased 27% from pre- to post-remediation cores measurements. The S_n estimated from the post-remediation partitioning tracer test was 31% greater than that from the pre-remediation partitioning tracer, which corresponds well to the increase determined with the core samples.

The strong vertical gradient in the center of the cell created by operation of the IWA appears to have mobilized and redistributed immiscible liquid within the cell. Comparisons between immiscible-liquid residual for the three extraction wells showed relatively uniform distribution for the pre-remediation study and a nonuniform distribution for the post study. During operation of the recirculation well, large quantities of floating immiscible liquid appeared in all wells and the shallowest MLSs. More than 7 L of immiscible liquid was collected from wells 2252 and 2253. This redistribution is most likely responsible for the contaminant-mass increases measured in the post-cores and post-partitioning tracer tests. Thus, while contaminant mass was removed with operation of the IWA, this amount is negligible compared to the impact of the immiscible-liquid redistribution.

The purpose of the studies conducted at OU1 of Hill AFB was to compare the performances of several remediation technologies. Hence, the tests were designed to be conducted in a manner as similar as possible. While the design worked well for the horizontal flushing technologies, it was not optimal for the IWA system. Thus, we expect IWA systems may be more successful in systems for which their use is optimized.

Acknowledgments

This material is based upon funded by the U.S. E.P.A. under cooperative agreement no. CR-822024 to The University of Arizona with funds provided through the Strategic Environmental Research and Development Program (SERDP). This document has not been subject to agency review; mention of trade names or commercial products does not constitute endorsement or recommendation for use. We would like to thank the many individuals at the U. S. Air Force, the U. S. E.P.A., and the University of Arizona who made this study possible. In particular Jon Ginn of USAF, Lynn Wood of the USEPA, and John E. McCray and Ken Bryan of the University of Arizona.

Literature Cited

(1)   Herrling, B.; Stamm J.; Buermann W. In *Proceedings International Symposium on In Situ and On-site Bioreclamation*; Hinchee, R.; Olfenbuttel R., Ed., Butterworth-Hienemann Boston, **1991**.

(2)   Herrling, B.; Stamm J. In *Proceedings Computational Methods in Water Resources IX Vol. 1. Numerical methods in water resources*; Russell R.; Ewing R.; Brebbia C.; Gray W.; Pinder G., Ed., Computational Mechanics Publication, Boston, **1992**.

(3)   IEG Technologies Corporation, Charlotte, **1998**.

(4)   Knox, R. C.; Sabatini, D. A.; Harwell, J. H.; Brown, R. E.; West, C. C.; Blaha, F.; Griffith, C. *Ground Water*. **1997**, *35*, 948.

(5)   Sabatini, D. A.; Knox, R. C.; Harwell, J. H.; Soerens, T.; Chen, L.; Brown, R. E.; West, C. C. *Ground Water*. **1997**, *35*, 954.

(6)   Coyle, J. A.; Borchers, H. J.; Miltner, R. J. EPA/600/2-88/020, USEPA, Cincinnati, **1988** *March*.

(7)   Herrling, B.; Buermann W. In *Computational Methods in Subsurface Hydrology;* Gambolati G.; Rinaldo A.; Brebbia C. A.; Gray W. G.; Pinder G. F., Ed., Spring-Verlag, New York, **1990**, 299-304.

(8)   Herrling, B.; Stamm J.; Alesi E. J.; Bott-Breuning G.; Diekmann S. In *In situ bioremediation of groundwater containing hydrocarbons, pesticides, or nitrates using vertical circulation flows*; Hinchee R., Ed., Butterworth-Hienemann, Boston, **1993**.

(9)   Bedient, P., Holder, A. W., Enfield, C. G., Wood, A. L., in this issue.

(10)   McCray, J. E., Brusseau, M. L., *Environ. Sci. and Tech.* **1998**, *32*, 1285.

(11)   Cooke, C. E. Jr., Method of determining residual oil saturation, *U. S. Patent 3,590,923*, U. S. Pat. Off., Washington, D. C., **1971**.

(12)   Tang, J. S., *SPE Formulation Evaluation*. **1995,** *March*, 33.

(13)   Jin, M.; Delshad, M.; Dwarakanath V.; McKinney D. C.; Pope G. A.; Sepehrnoori K.; Tilburg C. E., *Water Resour. Res.* **1995**, *31*, 1201.

(14)   Nelson, N. T.; Brusseau M. L., *Environ. Sci. and Tech.* **1996**, *30*, 2859.

(15)   Brusseau M. L., Nelson, N. T.; Cain, R. B. In this issue.

(16)   Gierke, J. S., Wojick, C. L., Hutzler, N. J. In the issue.

(17)   Montgomery, J. H., *Groundwater Chemicals Desk Reference, 2nd Ed.* Lewis Publishers, **1996**.

Chapter 13

Groundwater Remediation of Chromium Using Zero-Valent Iron in a Permeable Reactive Barrier

Robert W. Puls[1], Robert M. Powell[2], Cynthia J. Paul[1], and David Blowes[3]

[1]National Risk Management Research Laboratory, Robert S. Kerr Environmental Research Center, U.S. Environmental Protection Agency, Ada, OK 74820
[2]Powell & Associates Science Services, 8310 Lodge Haven, Las Vegas, NV 89123
[3]Department of Earth Sciences, University of Waterloo, Ontario N2L 3G1, Canada

A series of laboratory experiments were performed to elucidate the chromium transformation and precipitation reactions caused by the corrosion of zero-valent iron in water-based systems. Reaction rates were determined for chromate reduction in the presence of different types of iron and in systems with iron mixed with aquifer materials. Various geochemical parameters were measured to confirm the proposed reactions. Laboratory experiments were scaled up to pilot and full-scale field demonstrations. Intensive geochemical sampling in the field tests corroborate laboratory results and successfully demonstrate the effectiveness of this innovative *in situ* approach to remediate chromate-contaminated ground water using a permeable reactive barrier composed of zero-valent iron.

A great deal of recent interest has been focused on *in situ* techniques for treating contaminated ground water. One of the most promising of these techniques is the use of zero-valent metals in permeable reactive subsurface barriers for intercepting and remediating contaminant plumes. It has been shown that zero-valent iron (Fe^0) is effective for reductively dehalogenating halogenated hydrocarbons, such as trichloroethene (TCE). It can also reduce chromium from Cr(VI) to Cr(III), causing it to precipitate as an immobile and non-toxic hydroxide solid phase under most environmental conditions. Although these processes have been attributed to corrosion of the iron and cathodic reduction of the contaminants, there was little understanding of the reduction and precipitation transformation mechanisms. This research was implemented to develop an understanding of these processes and effects.

Purpose and Objectives

This research had a number of specific objectives, but its overall purpose was to

182

determine whether the use of a subsurface permeable reactive barrier (PRB) of zero-valent iron (Fe⁰) would remediate a chromate plume at the U.S. Coast Guard (USCG) Air Support Center in Elizabeth City, North Carolina. Although not a major focus of this study, it was also expected that the iron PRB would reduce or eliminate the overlapping plume of TCE that would intercept the wall (*1*).

It is known that in aqueous systems the distribution of chromium forms is strongly dependent on pH and redox potential (Eh) (*2, 3*). It occurs in two stable oxidation states in the subsurface environment, Cr(VI), or chromate, and Cr(III) (*4*). The chromate forms are of greatest concern due to their toxic and carcinogenic properties and their greatly increased subsurface mobility compared to the relatively immobile and nontoxic Cr(III) and its species. The U.S. EPA has set the drinking water MCL (maximum contaminant level) for chromium at 100 μg/L.

When reduced from Cr(VI) to Cr(III), Cr is removed from solution as $Cr(OH)_3$, a precipitated phase or, when dissolved iron is present, potentially as a chromium-iron hydroxide solid solution $(Cr_x, Fe_{1-x})(OH)_3(ss)$. The formation of this solid solution is desirable for remediation, because precipitated Cr is less soluble in this form (*5*). In addition, the removal of Cr from solution had been demonstrated to occur in the presence of zero-valent iron (*6*).

The specific objectives of the laboratory studies were:

- Determination of the rates of reactions for chromate reduction and precipitation by Fe⁰
- Developing an understanding of the reaction mechanisms
- Understanding the effects of native geochemistry on these rates and mechanisms
- Optimizing remediation of the Elizabeth City site with respect to the type(s) of Fe⁰

The objectives of the field studies included:

- Confirmation of effectiveness of Fe⁰ to remediate chromium-contaminated ground water *in situ* using permeable reactive barriers and
- Evaluation of the effectiveness of barrier installation methods.

To accomplish these goals, a number of laboratory experiments were carried out, along with pilot-scale and full-scale field studies at the Elizabeth City site.

Study Site

The field site is located at the USCG Air Support Center near Elizabeth City, North Carolina, about 100 km south of Norfolk, Virginia and 60 km inland from the Outer Banks region of North Carolina. The base is located on the southern bank of the Pasquotank River, about 5 km southeast of Elizabeth City. Hangar 79, which is located 60 meters south of the river (Figure 1), contained a chrome plating shop that operated for more than 30 years. Acidic chromium wastes and associated organic solvents were unknowingly discharged through a hole in the concrete floor of the plating shop. These wastes infiltrated the soils and the underlying aquifer immediately below the shop's foundation resulting in per cent levels of chromium in the soils (7).

The site geology has been described in detail elsewhere (7), but essentially consists of typical Atlantic coastal plain sediments, characterized by complex and variable sequences of surficial sands, silts and clays. Ground-water flow velocity is extremely variable with depth, with a highly conductive layer at roughly 4.5 to 6.5 meters below ground surface. This layer coincides with the highest aqueous concentrations of chromate and chlorinated organic compounds. The ground water table ranges from 1.5 to 2.0 m below ground surface.

Materials and Methods

Chromate Laboratory Studies. The experimental designs, materials setups and evaluations of the chromate laboratory studies have been extensively described elsewhere (8-12), therefore they will be summarized here without extensive detail or the presentation of supporting information, such as statistical analyses.

Iron Metals. Scrap iron filings from Ada Iron and Metal (AI&M, Ada Oklahoma), were the most comprehensively studied of the Fe^0-bearing metals and by far the most reactive for reducing chromate. This Fe was not pure and was visually observed to be partially oxidized. Cast iron metal chips were obtained from Master Builder's Supply (MBS, Streetsboro, Ohio). The metals were repeatedly washed with 7 mM NaCl in deionized water to remove any residual light hydrocarbons associated with the iron and freeze-dried. Chemical analysis showed that both could be classified as low-grade steels, with the AI&M consisting of 92.9% Fe and the MBS 90.1% Fe. The AI&M contained 0.75% carbon while the MBS was 2.98% carbon. Much of the MBS carbon seemed to be present on the surface of the chips as graphitic coatings and inclusions. BET (Brunauer, Emmett,Teller) (13) surface areas were 8.26 m^2/g for the AI&M and 1.10 m^2/g for the MBS.

Aquifer Materials and Water. The aquifer material used for most experiments was from the USCG Air Support Center at Elizabeth City. The site geology is Atlantic coastal plain sediment consisting of surficial sand, silt and clay. Core material from Otis Air Force Base, Cape Cod, Massachusetts, was also used to compare two distinct depositional environments. The Cape Cod sediment is composed of Pleistocene glacial deposits of sand and gravel with trace quantities of silt and clay sized particles. A standard silica sand (U.S. Standard Nos.10-20 mesh fraction) was also utilized but not characterized. A simulated aquifer water (SAW) was used which reflected the major ionic composition of the Elizabeth City site ($CaSO_4$ = 1mM, NaCl = 7 mM, pH = 6.5).

Stirred Batch Reactor Experiments. Some kinetics experiments were performed in stirred batch reactor (SBR) systems to determine Cr reaction kinetics and to evaluate changes in Eh and pH relative to these kinetics. No attempt was made to completely exclude oxygen in these experiments, although a humidified N_2 blanket was constantly maintained above the SBR. Oxidation-reduction potential (Eh), pH, and system temperature were continuously monitored throughout the experiments. Chromate reduction experiments were performed both in systems containing aquifer material and in systems containing only iron as an initial solid phase. In the experiments with aquifer material a solid : solution ratio of 1.5 : 3.5 (150 g aquifer material : 350 g Elizabeth City SAW solution) was utilized based on preliminary scoping studies, previous experience and published guidance on adsorption procedures (14).

After allowing a period of approximately 48 hours for the systems (solution or solution plus aquifer material) to approach equilibrium with respect to pH and Eh, 8 grams of the Al&M Fe filings were added to each reaction vessel. After a second equilibration period of approximately 60 hours, additions were made from a stock K_2CrO_4 solution, resulting in initial aqueous CrO_4^{2-} concentrations in the reactor systems ranging from 136 to 156 mg/L. Baseline samples were withdrawn from the SBRs prior to the chromate addition as well as periodic withdrawals after the addition. All samples were immediately filtered through pre-weighed 0.2 μm pore diameter Gelman nylon membrane filters. An aliquot of the filtrate was removed for sulfate and chloride analysis by capillary electrophoresis. The remaining filtrate was acidified to pH < 2.0 with ultra-pure nitric acid and analyzed by inductively-coupled plasma emission spectroscopy (ICP). The solids and filters were dried at 70°C and submitted for complete digestion and ICP analysis.

Shaken Batch Experiments. To evaluate effects of the variables aquifer material type, solid:solution ratio, iron mass and type, and SO_4^{2-} concentration, shaken batch bottle (SBB) experiments were used. Such batch experiments are useful in differentiating effects from multiple variables in a relatively efficient experimental fashion. A pseudo-factorial design was employed to reduce the number of replicates and repeated variables. These experiments were done in shaken Oak Ridge polyallomer centrifuge tubes in a N_2-filled glovebox. They were set up with 5 concentrations or masses of each variable, in duplicate.

Tubes of SAW and aquifer material were secured horizontally on a Thermolyne Maxi-Mix III rotating shaker at 300 RPM and equilibrated for 24 hours. Eh and pH were then measured in the glovebox, Fe metal added, and the samples returned to the shaker. After 48 hours, Eh and pH were again measured and CrO_4^{2-} added ($t_0 = 0$) to give a concentration of 126 mg Cr/L. The tubes were sampled in the glovebox at t = 20 minutes, 2, 7, 24, 48, and 96 hours. Samples were immediately filtered and analyzed as given previously for the SBRs.

Pilot-Scale Field Test. A pilot-scale field test was initiated at the USCG site in November 1994 by researchers from the National Risk Management Research Laboratory of the USEPA. It was the first such test to evaluate the use of zero-valent iron to remediate a chromium-contaminated aquifer.

Reactive Mixture. Based on results of the laboratory studies, zero-valent iron was mixed with aquifer material and a coarse sand to evaluate the effectiveness of the iron to remediate dissolved chromate at the Elizabeth City site. Two sources of iron were mixed and used in the field test. Low-grade steel waste stock, obtained from Ada Iron and Metal (Al&M, Ada, OK), was turned on a lathe (without use of cutting oils) using diamond bits to produce 200 L of turnings. The other iron material was obtained from Master Builder's Supply (MBS, Streetsboro, Ohio). The latter material was primarily in the 0.1 to 2 mm size range while the former was primarily in the 1 to 10 mm size range. The MBS iron exhibited greater total sulfur and carbon (as graphite) content than the Al&M iron.

Table I. Physical specifications for mixture used in the reactive barrier wall.

Mixture Components	% Vol.	% Wt.	Particle Size (mm)	Surf. Area (m²/g)
MBS Fe	25	29	0.2-4.0	1.1
Al&M Fe	25	33	1-15	1.4
E.C.Aquifer	25	19	<0.1	5.8
Gravel-Sand	25	19	1-4	<1

The aquifer material used was native to the site and was added to the mix from auger cuttings. A coarse, uniform, washed sand (3/16 x 10 mesh) was also added to the mixture for increased permeability within the iron cylinders. A summary of physical characterization data for the mixture is presented in Table I. The four materials were were mixed in equal volumes, on-site, and poured through emplaced 16 cm i.d. hollow stem augers that were then withdrawn. The estimated diameter of the emplaced cylinders of mix was 20 cm and they were installed from approximately 3 to 8 m below

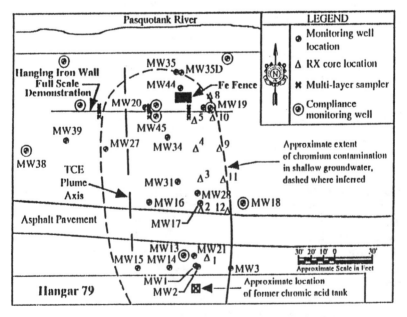

Figure 1. USCG site map showing plume locations, iron fence pilot test location and full-scale iron wall location.

ground surface. A total of 21 such cylinders were installed in three rows in that portion of the site labeled "Fe Fence" in Figure 1. Some of the boreholes were offset relative to the previous row(s) to evaluate the effects of spacing.

Monitoring Network. Twenty-four monitoring wells were installed within the approximately 5.5 m^2 treatment zone, in addition to four up-gradient reference wells and one downgradient well. Most are 1.9 cm i.d. polyvinyl chloride (PVC) wells with 30 or 45 cm long screens which are completed between 4.2 and 6.0 m below ground surface. In addition to the permanent well sampling points, temporary sampling points were utilized to increase the spatial resolution of the data. These were obtained using a Geoprobe® and peristaltic pump.

Field Analyses. Analyses for chromate, ferrous iron, dissolved sulfide, dissolved oxygen (DO), specific conductance, temperature, pH, redox potential (Eh), alkalinity, and turbidity were performed in the field. The species Cr(VI), S^{2-}, and Fe(II) were analyzed colorimetrically with a UV/VIS spectrophotometer (Hach DR2010). Dissolved oxygen (DO) was measured using a CHEMets® colorimetric test kit which utilizes a rhodazine-D colorimetric technique and in some cases using a ORION model 810 DO meter with ORION 81010 DO electrode. Conductivity and temperature measurements were made using a conductivity probe and meter (ORION Conductivity meter, model 128 and/or model 135). Eh and pH measurements were made using platinum redox and glass bulb pH electrodes in a sealed flow-through cell. Alkalinity measurements were made by titration with standardized H_2SO_4 acid using a Hach® digital titrator. Turbidity was measured with a Hach® turbidimeter.

Sample Collection. Ground water samples were collected with a peristaltic pump using low-flow purging and sampling techniques (*15, 16*). Samples were taken behind the pump head for inorganic analytes and before the pump head for organic analytes, to minimize losses of volatiles and gases. All samples were collected following equilibration of water quality parameters (DO, pH, Eh, specific conductance). Equilibration of water quality parameters was defined as three successive readings within ±10% for DO and turbidity, ±3% for specific conductance, ±10 mv for Eh, and ±0.1 for pH. Filtered and unfiltered samples were taken for metals and cation analysis and acidified to pH < 2 with ultra-pure concentrated nitric acid.

Laboratory Analyses. Total metals were analyzed using a Jarrell-Ash Model 975 Inductively Coupled Plasma (ICP) spectrometer. Anion samples were unfiltered and unacidified, and analyses were performed using ion chromatography (Dionex DX-300) or using Waters capillary electrophoresis (Waters Method N-601). Solids have been analyzed using scanning electron microscopy with energy dispersive x-ray (SEM-EDS) and secondary ion mass spectrometry (SIMS).

Full-Scale Demonstration. A full-scale PRB demonstration to remediate ground water contaminated with chromate and chlorinated organic compounds was initiated at the USCG site by researchers from USEPA's National Risk Management Research Laboratory and the University of Waterloo in 1995. A continuous wall composed of

100% zero-valent Peerless iron (Peerless Metal Powders, Inc.) was installed in June, 1996, using a trencher that was capable of installing the granular iron to a depth of 8 m. The trenched wall was approximately 0.6 m thick and about 60 m long, its location on-site indicated in Figure 1. It was installed with the trencher in less than 8 hours, whereby native soil and aquifer sediment was removed and the iron simultaneously emplaced in one continuous excavation and fill operation. Despite favorable results obtained in the laboratory and the field-pilot test using a mixture of iron plus aquifer sediment, 100% iron was used in the full-scale demonstration because of logistical and handling difficulties associated with installing 500 tons of iron with the trencher.

Results and Discussion

The results of the laboratory studies have been detailed in several publications (listed in Methods and Materials, above), and will be summarized here.

Reaction Rates for CrO_4^{2-} Reduction. Results from stirred batch reactor experiments indicated the loss of chromate from solution and its increase in the filtered solids. Table II provides solution data for two of the SBRs, one of which contained no aquifer material. With no aquifer material present, chromate-Cr fell from an initial concentration of about 140 mg/L to a concentration of 26 mg/L during a 146 hour period following its injection into the system. In contrast, when the aquifer material was present the chromate was reduced below ICP detection limits (0.0024 mg/L) during a 60 hour period. The pseudo-first order reaction rates and resultant half-lives ($t_{1/2}$) are also provided in Table II.

The presence of Elizabeth City aquifer materials in the reactors was found to buffer both the Eh and the pH of the reacting systems (9, 10, 12). Increases in pH were much larger in the systems without aquifer solids, rising as high as pH > 10 when the Fe filings were added. In the systems containing Elizabeth City aquifer solids, the pH did not exceed pH = 8.0. This pH buffering by the aquifer material is significant since it is known that CrO_4^{2-} reduction occurs more readily at lower pH.

In addition to having a lower pH, approximately 50% of the chromate was removed from solution before the first sampling event (< 1 min after chromate introduction) in the system containing Elizabeth City aquifer material via either reduction and precipitation or adsorption. This was not due solely to Cr sorption on the natural aquifer material. The Langmuir sorption capacity term was determined (14) and indicated that only 6.7 % of the Cr could be sorbed by the aquifer material in the reactor if the pH was 6.5 (natural Elizabeth City aquifer pH). At the reactor system pH ≈ 7.5 the aquifer materials would have an even higher net negative charge, decreasing adsorption of negatively-charged chromate. However, due to corrosion of the iron filings, the reactor system also contained a large quantity of solid phase iron hydroxides/oxyhydroxides which could greatly increase the chromate sorption capacity of the overall system (17, 18). However, the other experiment (no aquifer material), contained the same initial mass of iron filings and no instantaneous loss of chromate was seen. This could be considered evidence that chromate reduction was favored over sorption, if it were not for the difference in pH of the two systems. The iron corrosion phase most likely to initially form is amorphous $Fe(OH)_3$, which has a pH_{zpc} of 8.5.

Table II.Chromate pseudo-first order reduction rates and half-lives for two stirred batch reactors, each containing 8 g of AI&M Fe filings in 350 ml of simulated Elizabeth City water.

EC Aquifer Material	Reaction pH	C^{final} mg/L	Reaction Rate hr^{-1}	Half-Life $(t_{1/2})$, hr
None	≥8.7	26	0.010	67.9
150 g	7.5	<0.002	0.064*	10.8*

*Rate and half-life for this experiment exclude the region of immediate Cr disappearance.

Therefore, at pH = 7.5, a large surface area having a net positive charge is present, creating the potential for significant sorption of CrO_4^{2-}. The pH range 8 to 10 is significantly higher than the pH_{zpc} of the $Fe(OH)_3$. This greatly reduces the number of positively charged surface sites available for chromate adsorption. Other potential iron oxide phases present on the iron filings surface (e.g., Fe_2O_3, Fe_3O_4) have similar pH_{zpc}'s (7 to 8.6) and if present would have similar effects with respect to adsorption behavior.

While organic carbon could also account for chromate reduction, the organic carbon content of the Elizabeth City aquifer material (like most aquifer materials) is quite low (≤0.10 %). Earlier studies at the site have shown surface soil organic carbon to be capable of reducing chromate, but little was observed for the aquifer sediments (7).

Therefore the Elizabeth City aquifer material affected the zero-valent iron system in several ways that could account for the immediate disappearance of CrO_4^{2-} from solution and increase the Cr reduction rates:

(1) Lowering the reaction pH; it is well known that corrosion reactions occur more rapidly at low pH.

(2) Lowering the pH of the system below the pH_{zpc} of $Fe(OH)_3$, allowing adsorption of chromate to these oxyhydroxides.

(3) A relatively small amount of adsorption (≤ 6.7%) of chromate to the Elizabeth City aquifer material itself.

(4) Increasing the number of particles that could promote nucleation prior to precipitation, and

(5) Increasing the surface area possessing reactive sites that could adsorb intermediate Cr^{3+} species (e.g., $Cr(OH)_n^{(3-n)}$) prior to complete formation of the hydroxide precipitate.

Exploratory data analyses of the shaken batch bottle experiments (8, 10) determined that the most important factor affecting the removal of chromate from the system was the quantity of iron present for a given aqueous system volume. Also, for a given mass of iron, the rate of removal was more rapid at increased solid:solution ratios of Elizabeth City aquifer material, confirming that these materials increased the rates of Cr removal from solution. The best kinetic fits to the data were also pseudo-first order with respect to the disappearance of Cr from the solutions, as it was for the SBRs.The scrap AI&M iron filings had shorter reaction half-lives than the MBS iron. Table III provides information on these rates per square meter of Fe surface area from some of the batch experiments.

Table III. Chromate pseudo-first order reduction rates (K) per square meter of Fe surface in the absence and presence of varying quantities of Elizabeth City aquifer material in 26 ml of solution, along with the ratios of the rates for AI&M Fe versus MBS Fe.

EC Aquifer Material, g	$K_{AI\&M}$ hr^{-1} $(m^2)^{-1}$	K_{MBS} hr^{-1} $(m^2)^{-1}$	$K_{AI\&M}/K_{MBS}$
0.00	4.41E-2	7.59E-3	5.81
0.05	2.09E-2	7.05E-3	2.96
0.50	3.96E-2	6.29E-3	6.30
5.00	3.18E-1	7.67E-3	41.4
20.00	3.88E-1	1.20E-2	32.5

Experiments using aquifer materials from Cape Cod yielded much less enhancement of chromate reduction than the Elizabeth City material, while experiments containing a commercial silica sand showed even less enhancement. Table IV provides chromate half-life results with these materials.

Pilot-Scale Field Study Results. Table V summarizes the changes in chromium concentration over time for wells GMP1-2 and 1-3 (upgradient reference wells), and downgradient wells F7, and F9, following initial "fence" emplacement September 13-16, 1994. Initial conditions are indicated by the June, 1994 sampling data. Well screen depth intervals are also listed in Table V. For the upgradient wells, chromium concentrations are relatively stable and range from 1 to 3 mg/L depending upon depth. immediately behind iron cylinders. In both cases, chromium concentrations immediately declined to less than 0.01 mg/L, and iron increased substantially. Almost all of the iron All iron detected in the upgradient wells was colloidal; there was no detectable ferrous

Table IV. First order chromate half-lives in shaken batch bottle experiments with differing aquifer materials. Each bottle contained 0.25 g of AI&M Fe, 26 ml of simulated aquifer water (pH 6.5, 0.86 mM $CaSO_4$, 7 mM NaCl), and when present, 9 g of the aquifer material.

Aquifer Material	Chromate $t_{1/2}$, hours
Elizabeth City + Fe	12
Cape Cod + Fe	26
Silica Sand + Fe	119
Fe with no aquifer material	>1000

Table V. Monitoring well data in the iron fence area.

Well	Screen Interval Depth (m)	Initial (prior to iron emplacement) Concentration, Cr (mg/L)	Final (following Iron emplacement) Concentration, Cr (mg/L)
F7	4.8-5.2	1.7	< 0.01
F9	4.7-5.1	1.7	< 0.01
GMP1-2	4.9-5.2	1.7	1.1
GMP1-3	5.4-5.7	0.7	3

iron, and there was no increase in iron concentration. Wells F7 and F9 were located detected was ferrous iron. Changes in geochemistry over this same time period for these same (F series) wells was also monitored. Variation in effectiveness of contaminant removal from the aqueous phase was directly correlated with changes in geochemistry and position of monitoring points relative to the iron cylinders. Wells located within or downgradient of the "fence posts" show reduction of chromate to below detection limits (< 0.01 mg/L). Ferrous iron increased to 20 mg/L, dissolved hydrogen increased from less than 10 nMol to more than 1000 nMol, Eh decreased as low as -100 mV, pH increased, DO was consumed, and sulfides were detected in both the aqueous and solid phases. These geochemical changes were identical to prior laboratory observations by Powell et al. (10) and are consistent with the following reactions:

$$Fe^o + 2H_2O \rightarrow Fe^{2+} + H_2 + 2OH^-$$

$$Fe^o + CrO_4{}^{2-} + 4H_2O \rightarrow (Fe_xCr_{1-x})(OH)_3 + 5OH^-$$

The geochemical changes occurring further downgradient of the iron fence with respect to an individual "fence post" of the iron mixture exhibit a range of progressively decreasing redox zones. In addition to an increase in pH, decrease in Eh, consumption of dissolved oxygen and increase in alkalinity, there is the generation of detectable sulfides and small amounts of ferrous iron in the ground water.

Sulfides are not detected beyond 0.3 m downgradient from an iron cylinder and ferrous iron persists only for about 1.5 m downgradient. Sulfides were not detected in the laboratory studies. Likely phases which may form from the iron corrosion and natural water/sediment/microbial interaction are metal sulfides, siderite and iron oxyhydroxides. Iron sulfides and a mixed iron-chromium oxide or hydroxide phase have been observed using SIMS and SEM-EDS; siderite has not been identified.

Field and laboratory results indicate almost instantaneous removal of chromium from the aqueous phase, and chemical extraction together with surface analytical techniques confirm that this is through reduction and precipitation reactions forming a very insoluble mixed chromium-iron hydroxide phase. Field geochemical changes observed as a result of iron mixture emplacement in the aquifer are consistent with those observed in the laboratory experiments.

Full-Scale Implementation Results. To date there have been two years of post-installation performance monitoring for the full-scale implementation at the Elizabeth City site. For all but one quarterly sampling event, 15 multi-level samplers (7 to 11 sample ports per sampler) and 9 to 10 compliance (5-cm PVC) wells have been sampled. In addition to onsite sampling of the full suite of geochemical indicator parameters listed in the site work plan, samples have been collected for laboratory analysis of the following constituents: TCE, cis-dichloroethene, vinyl chloride, ethane, ethene, methane, major anions, and metals using the same procedures as in the pilot-scale test. In addition, numerous vertical and angle cores have been collected to examine changes to the iron surface over time and to evaluate the formation of secondary precipitates that may affect wall performance over time. Electrical conductivity profiles have also been conducted upgradient and within the wall to delineate and confirm iron emplacement and treatment.

Upgradient conductivity profiles correlate well with major anion chemical distribution and chromate distributions vertically. Profiles within the wall indicate similar correlations and provide assessment of the granular iron distribution as well. The cores collected thus far indicate the iron was emplaced approximately as expected depth-wise, but it appears to be somewhat less thick than the 60 cm design. Coring was done vertically (perpendicular to ground surface) and on an angle (30°). The former method provided continuous vertical iron cores, while the latter provided a transverse core through the wall with the aquifer-iron interfaces intact (front and back of the wall). These cores continue to be under study. Inorganic carbon contents increase at the upgradient aquifer sediment-iron interface and decrease within the wall moving downgradient, reaching background levels within 10 cm downgradient from the downgradient iron-aquifer sediment interface. Total inorganic carbon content has increased over time within the wall. Water levels indicate little difference (< 0.1 m) between wells completed and screened at similar depths upgradient and downgradient of the wall, indicating that the wall continues to effectively function as a "permeable" reactive barrier.

Sampling results for chromate indicate that all chromate was removed from the ground water within the first 15 cm into the wall, as expected. No chromate is detected downgradient of the wall either in the multilayer samplers or in the 5-cm compliance wells immediately behind the wall. The wall appears to have successfully resolved the chromate contamination of the aquifer in terms of the diffuse plume migrating towards the river! This is a significant accomplishment and the first field demonstration of this technology to remediate a chromate-contaminated aquifer to regulatory standards.

Results of geochemical sampling onsite indicate that iron corrosion is proceeding within the wall. There are significant reductions in Eh (generally < -400 mV), increase in pH (> 9), absence of DO, and decrease in alkalinity. Downgradient of the wall (1.5 m), pH returns to near neutral and Eh is quite variable with depth. Over time there have been indications that a redox front is slowly migrating downgradient within the first few meters of the wall. Table VI shows comparisons between measured geochemical parameters for the aquifer, the iron "fence" and the full-scale wall. In both cases,

Table VI. Summary of geochemical monitoring parameters for aquifer, iron "fence", and full-scale iron wall.

Monitored Parameter	Aquifer	Continuous Iron Wall	Iron Fence
Cr(VI)	1-5 mg/L	<0.01 mg/L	<0.01 mg/L
Fe(II)	<0.2 mg/L	<0.2-15 mg/L	2-20 mg/L
Eh	300 - 600 mV	-300 to - 550 mV	-100 to +200 mV
pH	5.7 - 6.5	9.1-10.7	6.6-9.0
DO	0.2 - 2.0	<0.2 mg/L	<0.2 mg/L
dissolved H_2	< 10 nMol	> 1000 nMol	NA*
SO_4	5 - 80 mg/L	< 5 mg/L	< 5 mg/L
Alkalinity	20 - 100 mg/L ($CaCO_3$)	decrease	variable
Precipitates on Iron Filings		iron oxides, hydroxides, oxyhydroxides; green rust (?)	iron oxides, hydroxides, oxyhydroxides; iron sulfides

* Dissolved hydrogen was not analyzed for in the pilot-scale test.

chromate was reduced and removed from the aqueous phase; a significant increase in ferrous iron, pH, and dissolved hydrogen was observed; and decreases in Eh, DO, and sulfate occurred. Alkalinity decreased in the continuous wall but this was not obvious in the "fence" data. Changes in pH and Eh in the "fence" data were not as dramatic as for the continuous wall due to the buffering effect of the aquifer sediments mixed with the iron in the "fence" installation. The continuous wall utilized 100% iron as the reactive media. Iron sulfides have been observed in the "fence" reaction zone mixture, but not identified to date on the iron samples recovered from the continuous wall. Green rust has tentatively been identified on iron filings recovered from the continuous wall. This would account for at least part of the sulfate removal in that system. Increases in alkalinity were observed in some sampling points in the "fence" test suggesting microbial-mediated reduction of sulfur presumably by sulfate reducing bacteria. It is likely that these bacteria were present within the fresh aquifer material mixed with the iron for that test and would explain the appearance of sulfides in that test and not in the full-scale wall where no aquifer materials were mixed with the iron. The pH buffering provided by the sediments would also have been favorable for enhanced microbial activity in the reaction zone mixture within the "fence" area.

194

Disclaimer

Although the research described in this article has been funded wholly or in part by the United States Environmental Protection Agency, it has not been subject to the Agency's peer and administrative review and therefore may not necessarily reflect the views of the Agency and no official endorsement may be inferred.

Literature Cited

1.   Gillham, R. W.; O'Hannesin, S. F. *Ground Water* , **1994**, *32*, 958.
2.   Robertson, F. N. *Ground Water*, **1975**, *13*, 516.
3.   Sloof, R. F. M. J.; Cleven, J. A.; Van der Poel, J.; Van der Poel, P. *Integrated criteria document: Chromium*; RIVM-710401002; National Institute of Public Health and Environmental Protection; Bilthoven, The Netherlands, **1990**.
4.   McLean, J. E.; Bledsoe, B. E. *Behavior of Metals in Soils*, U.S. Environmental Protection Agency, Robert S. Kerr Environmental Research Laboratory: Ada, OK, **1992**, EPA/540/S-92/018.
5.   Rai, D.; Zachara, J. M. *Geochemical behavior of chromium species*. EPRI EA4544. Electric Power Research Institute. Palo Alto, CA, **1986**.
6.   Blowes, D. W.; Ptacek, C. J. Subsurface Restoration conference, Third International Conference on Ground Water Research; National Center for Ground Water Research: Dallas, TX, **1992**; pp. 214-216.
7.   Puls, R.W.; D.A. Clark; C.J. Paul; J. Vardy. *J. Soil Contam*. **1994**, *3*, 203.
8.   Powell, R. M. Geochemical Effects on Subsurface Chromate Reduction and Remediation Using the Thermodynamic Instability of Iron. M. S. Thesis, The University of Oklahoma, Norman, OK, **1994**.
9.   Powell, R. M.; Puls, R. W.; Paul, C. J.   In *Innovative Solutions for Contaminated Site Management*, Water Environment Federation, Miami, Florida, March 6-9, **1994**, pp. 485-496.
10.   Powell, R. M.; Puls, R. W.; Hightower, S. K.; Sabatini, D. A. *Environ. Sci. Technol*. **1995(a)**, *29*, 1913.
11.   Powell, R. M.; Puls, R. W.; Hightower, S. K.; Clark, D. A. In 209th ACS National Meeting, American Chemical Society, Division of Environmental Chemistry; Anaheim, CA, April 2-7, **1995(b)**, pp. 784-787.
12.   Powell, R. M.; Puls, R. W. *Environ. Sci. Technol*. **1997**, 31, 2244.
13.   Brunauer, S.; Emmett, P H.; Teller. E. *J. Am. Chem. Soc*. **1938**, *60*, 309.
14.   Roy, W. R.; Krapac, I.G.; Chou, S.F.J.; Griffin, R.A. **1991**. USEPA Technical Resource Document, EPA/530-SW-87-006-F.
15.   Puls, R. W.; Powell, R. M. *Ground Water Monitoring Review*. **1992**, *12*,167.
16.   Powell, R. M.; Puls, R. W. *Pollution Engineering*. **1997**, *29*, 50.
17.   James, B. R.; Bartlett, R. J. *J. Environ. Qual*. **1983**, *12*, 177.
18.   Zachara, J. M.; Girvin, D. C.; Schmidt, R. L.; Resch, C. T. *Environ. Sci. Technol*. **1987**, *21*, 589.

INNOVATIVE SITE
CHARACTERIZATION METHODS

Chapter 14

Considerations for Innovative Remediation Technology Evaluation Sampling Plans

Michael J. Barcelona and David R. Jaglowski

Department of Civil and Environmental Engineering, University of Michigan, Ann Arbor, MI 48109-2099

Field trials of innovative subsurface cleanup technologies require the use of integrated site characterization approaches to obtain critical design parameters, to evaluate pre-treatment contaminant distributions, and to assess process efficiency. This review focuses on the transition between typical levels of site and source characterization at host locations and the levels necessary to provide substantial proof of performance. Sampling plans should coordinate the collection of samples for aquifer property and contaminant measurements to minimize program costs and the disturbance of test site conditions. Performance monitoring approaches which utilize comparisons of cumulative mass removals or 'before-and-after' contaminant levels will require the use of proven site characterization methods and frequent sample replication. Replication of samples improves estimates of natural and experimental variability, which should enable more precise assessments of treatment efficiency and bolster the acceptance of evaluations by regulatory officials. Rapid field-based analytical approaches can play a major role in controlling technology evaluation project costs.

Current literature on hazardous waste site characterization demonstrates that the selection of sampling locations, sampling frequencies, and sample collection methods will significantly influence the results (*1-3*). Contaminant detection and risk assessment efforts have emphasized the use of ground-water monitoring well networks with minimal attention to the determination of contaminant source and aquifer property characteristics, yet these characteristics vary substantially from site to site (*4-6*). Landfills, industrial sites, and fuel or solvent spill source areas ranging from tens of hectares to less than 100 m^2 may produce dissolved contaminant plumes that extend tens of meters into the unsaturated and saturated zones. Aquifer transport properties (*e.g.*, hydrostratigraphy, porosity, hydraulic conductivity, and dispersivity) are known to vary semi-randomly or to be correlated with the scale of transport studied (*7-9*). When combined with the varying physical, chemical, and biological properties of contaminant mixtures, these conditions can lead to significant vertical and horizontal variability in chemical and microbial distributions.

Conventional networks of upgradient and downgradient monitoring wells with screened intervals of 1 to 5 meters may fail to capture the complexity of subsurface hydrogeochemistry (4). Well screens need to be designed and located to intercept actual chemical transport pathways and to define the spatial evolution of a plume as influenced by local hydrologic conditions. Misplacement or the misuse of long-screened monitoring wells will cause the nature of the treatment problem to remain undiagnosed or underestimated by as much as an order of magnitude (1). For example, the use of bailer sampling methods in long-screened wells can lead to order of magnitude errors in assessing dissolved-phase contaminant distributions due to vertical averaging within the screened interval (4-6,10). The use of discrete multilevel sampling wells or depth discrete sampling techniques in long-screened wells (e.g., micropurging devices or diffusion cells) may mitigate the problems associated with vertical averaging and provide a significantly enhanced representation of subsurface chemical distributions (11).

Non-aqueous phase liquids (NAPLs) and sorbed contaminants represent long-term sources of dissolved-phase contamination that are poorly resolved, if at all, using monitoring wells. Since multiple phases are frequently involved, it is almost always necessary to conduct careful aquifer sediment sampling and analyses (12), as well as conventional characterization measures (13). Thorough sediment sampling plans should incorporate an hypothesized 'history' of NAPL migration at the site. Fluctuations in the water table of unconfined aquifers will tend to distribute residual phases of NAPL less dense than water (LNAPL) well into the vadose zone and below the water table in a 'smear zone' that emerges after several hydrologic seasons. NAPL more dense than water (DNAPL) will tend to leave residual phases along any paths of downward migration in the aquifer. NAPLs in general can be expected to distribute residual phases within the unsaturated zone between a surface spill location and the water table that may act as a persistent source of dissolved-phase contamination in regions with significant amounts of infiltration (12).

It should be mentioned that complete resolution of the source characteristics may not be absolutely necessary in preliminary detection and assessment efforts. However, in order to implement field scale operations or pilot technology evaluations a thorough characterization is essential. In these cases, the goal is to demonstrate source or plume control via hydraulic isolation or contaminant removal, enabling management of long-term risks of exposure to humans and the environment. Removal or conversion operations must be evaluated by rigorous techniques to demonstrate quantitative mitigation of long-term risks before a site can be closed or a promising technology can be selected for scale-up.

Site Characterization for Technology Evaluations

Published protocols for the selection of hazardous waste treatment technologies too often provide 'decision-free', generic guidance on the selection of appropriate techniques in broad categories of subsurface settings (14-17). It should be stressed that in-situ site characterization needs to become more intensive when active treatments are applied to complex mixtures of contaminants in varied hydrogeologic settings, which is a situation that is commonly encountered in present technology evaluations. The effects of the selected technology on the physical, chemical, and biological properties of the subsurface should be very carefully considered in these situations.

Treatment evaluations require the determination of many site-specific characteristics that can be grouped according to the nature of the technology of choice. It may be assumed that knowledge of the extent and spatial distribution of contamination and hydrogeologic setting are common requirements for all treatment technology evaluations. Table I contains a listing of commonly examined site characteristics for field evaluations of either active (e.g., involving induced vapor or liquid movement) or passive (*e.g.*, intrinsic or augmented bioremediation) treatment technologies. All treatment technologies are active in that specific unit operation(s) will be superimposed on the dynamics of ongoing physical, chemical and biological processes. In this sense, site characterizations for biologically-based treatments are especially complex, requiring detailed sampling and analysis of the operative physical, chemical, and biological factors, all of which may be quite variable in space and time.

Table I. Site Characterization Needs for Treatment Technology Evaluations

Phase	Physical ——————⟶	Chemical ——————⟶	Biological
Vapor	Temperature Pressure	Gas Composition	Moisture Content Gas Composition

Aqueous	Hydraulic Head
	  Distribution
	Temperature
	Density
		pH, Alkalinity
		Major Cations/Anions
		Trace Chemical
		  Constituents
		  (Fe^{2+}, NO_2^-, $S^=$)
		Dissolved Gases
		Dissolved Contaminants
		Dissolved Organic
		  Carbon
		Redox Potential
			Electron Donors/
			  Acceptors
			Microbial Activity
			Microbial Biomass
			Biotransformation
			  Rates
			Inorganic and
			  Organic
			  Metabolites

Sediment	Grain-Size Distribution
	Density (Particle and
	  Bulk)
	Porosity
	Temperature
	Hydraulic Conductivity
	Specific Conductance
		Mineralogy
		Inorganic/Organic
		  Carbon Content
		Redox Capacities
		Adsorbed Contaminants
		NAPL Composition
		NAPL Saturation
		Water Saturation
			Electron Donors/
			  Acceptors
			Microbial Activity
			Microbial Biomass
			Biotransformation
			  Rates
			Inorganic and
			  Organic
			  Metabolites

The suitability of many technologies may be dependent upon spatially variable geochemical or geologic properties that cannot be easily modified. For example, the choice of chemical approaches such as co-solvent based NAPL solubilization or surfactant based NAPL mobilization would depend heavily on subsurface permeability, potential for sorption or emulsion formation, and major ion content *(17)*. Mobilization technologies are particularly susceptible to unexpected effects caused by variations in subsurface permeability that are not revealed by aquifer sediment sampling at five foot intervals as opposed to continuous cores. To control project costs while maximizing the usefulness of a site characterization, the sampling plan of each technology evaluation must be designed to emphasize the quantity and distribution of samples collected for evaluation of a proposed treatment's major design variables.

Thorough site characterization is particularly necessary in contaminated environments where natural geologic or geochemical equilibria may be altered drastically. For example, biological dehalogenation could be selected to treat a hypothetical acidic reducing ground-water plume containing dissolved chlorinated solvents. Treatment could occur after the plume's acidity has been neutralized by the carbonate content of the aquifer sediments. Alternatively, treatment could occur after the plume's acidity has been neutralized by the careful injection of an appropriate base, provided that injection problems (*e.g.*, mineral precipitation or biofouling) can be addressed to allow effective delivery of the reagent. In this case, knowledge of both physical and chemical aquifer characteristics would be essential to the selection of an appropriate neutralization technique and successful application of the biological treatment method.

Given the high costs involved in the selection and implementation of in-situ treatment approaches (*18*), considerable effort should be placed on the analysis and use of existing and preliminary site characterization data prior to the commencement of additional sampling for treatment technology evaluations. This data can either rule out the use of a potential technology or be integrated into the more detailed data collected to support a treatment evaluation project. Furthermore, substantial effort should be placed on maximizing the value of any additional sampling efforts. Multimedia sampling devices and drive point samplers can be used to collect large volumes of material that can be split to supply a variety of analyses while minimizing disturbance of the test site and sample collection costs. If available, 'hydraulic push' devices like the GeoProbe (GeoProbe Systems, Salina, KS) in combination with drive point sampling may be used to refine site characterization prior to the installation of permanent monitoring wells. These devices can also be used to install small (*i.e.*, ≤ 5 cm o.d.) monitoring wells while minimizing disturbance of the surrounding aquifer material in small or densely monitored test sites.

Phase Distribution of Contaminants and the Role of Rapid Field Analyses

It has been recognized in the last ten years that organic contaminant mixtures partition effectively between the ground water and sediments in aquifers (*19-22*). This is true even for volatile organic compounds (*e.g.*, benzene, perchloroethene, trichloroethene, etc.) that have relatively low water-sediment distribution coefficients based on the fraction of organic carbon in the solid phase. This realization was spurred by improvements in sediment sampling, field preservation techniques, and analytical methods. It has been demonstrated subsequently that half or more of the contaminant masses may reside on the aquifer sediments, rather than in the associated ground water (*20-23*).

Therefore, it has become increasingly more common for pre-treatment site characterization efforts to include both ground water and sediment sampling and analysis. Critical steps in mass estimation now include rapid field preservation of core materials, field and confirmatory laboratory analyses, as well as accurate determinations of particulate organic carbon. The latter determinations can be problematic since many aquifer sediments, especially clean sands, contain percent-level carbonate mineral phases that must be distinguished from organic carbon at levels generally less than one percent (*24,25*). Oily phase content can be estimated by determinations of total petroleum hydrocarbons (TPH) by various methods (*e.g.*, Petroflag, USEPA Method 418.1) and analyzed in detail via chromatography of gasoline or diesel range organics to enable profiling of the total sediment associated organic matter. The former may be an appropriate and much less expensive surrogate measure of contamination at sites containing complex mixtures. The latter will be necessary for regulatory acceptance and may be used to obtain estimates of contaminant mass, potential microbial substrate mass, and electron donor/acceptor requirements.

The collection and analysis of point samples from subsurface cores can be approached by a number of different methods. In one particularly interesting example, Sale and Applegate (*26,27*) used visual inspection of an oily phase residual on a large number of cores (*i.e.*, >1000) before and after a water flooding operation to recover creosote oils. This visual logging procedure was correlated with a smaller subset of analytical determinations of mobile and residual DNAPL to provide volumetrically averaged mass removal estimates. They found that the visual logging procedure was accurate within a factor of two. These results are encouraging since the cost associated with hundreds of specific organic compound determinations would be prohibitive for most technology research and development efforts, as well as many operational field trials. The appropriate use of a statistically correlated surrogate measurement may improve the quality of an evaluation's dataset, because the spatial variability of contaminants or organic carbon content within a field test site may be comparatively high. Regardless of the analytical methods used, it is critical to achieve appreciable levels of sample collection and replication.

Experimental Design Considerations

Guidelines for the conduct of innovative technology evaluations have evolved substantially in the past decade. Common to most of the existing protocols is the requirement of a careful experimental design for data collection and analysis (17,28-31). Designs vary substantially regarding the need for experimental controls, levels of quality assurance/control, and the degree of replication required during the sampling and analysis phases of site characterization or treatment process monitoring. The choice of appropriate control conditions with equivalent levels of sample replication will affect data comparability and the confidence levels associated with subsequent statistical judgements.

Some experimental designs call for the comparison of subsurface conditions over time within the same treatment zone (i.e., contrasting contaminant mass within a unit volume prior to and after the application of the treatment being evaluated). These evaluations require close attention to pre- and post-treatment contaminant mass distributions in both the sediments and ground water, corrected for cumulative mass removal by the produced fluids. Multi-media spatial and time-series analyses are among the more challenging environmental sampling and analysis problems since propagated or off-setting errors directly impact the comparability of 'before and after' results (32,33). These difficulties go well beyond the inherent problems associated with comparisons of small numerical differences between large numbers, comparisons of sample means from sub-populations with distinctly different variances, or comparisons of sub-populations containing correlated data. In particular, strict comparisons of three-dimensional solid phase sample results are statistically questionable, even if the samples are assumed to be normally distributed, unbiased and of stable variance structure (33).

The use of advanced geostatistical spatial interpolation techniques, such as kriging, has been limited by a comparatively large minimum sample size, typically 30 or more data points within the averaging plane, and/or a lack of spatial correlation in the raw or transformed data. The most rigorous applications of kriging techniques have been for surface soil investigations where large arrays of point data lie in the same plane (34-37), although kriging in three dimensions can be performed with extensive datasets. Considerable work remains to be done to better guide the selection of appropriate semivariograms which can provide an appropriate basis for the spatial interpolation and estimation of uncertainty associated with estimates of various geophysical properties from point sample data.

Regardless of the analysis technique applied, it is important to conduct pre- and post-treatment data analyses in as near to identical manners as possible. This should be built into the experimental design, particularly with respect to the selection of indicator compounds that may be differentially removed or transported during pilot treatment evaluations. The most commonly applied two-population statistical analyses are often based upon hypothesis testing using the null hypothesis, which states that no difference exists between the sample populations (32). Hypothesis testing, whether it is based upon null hypotheses or affirmative hypotheses, will fail to fulfill its intended analytical role if unrecognized 'artificial' differences are imposed upon the two populations by varying sampling techniques, analytical methods, etc. Practicing scientists and engineers are often compelled to make decisions based on 'less than adequate' highly variable datasets and such comparisons will continue to be made. It is important to recognize where gross errors enter sampling and analytical processes. These errors must be controlled during treatment evaluations so that meaningful confidence levels can be associated with the results. The following examples are provided to guide experimental design efforts towards these ends. The examples have been drawn primarily from recent work applicable to multi-media sampling and analysis in all phases of site characterization. They highlight design aspects that should be considered to minimize gross errors in subsurface cleanup technology evaluations.

Pilot Treatment Evaluation Example – Subsurface Variability

An example that highlights observed levels of spatial variability and the necessity of adequate sample replication can be drawn from the results of a recently completed subsurface bioaugmentation experiment (*38*). A portion of the saturated glacial outwash sand deposits at this site had become contaminated with carbon tetrachloride (CCl$_4$) and chloroform (CHCl$_3$), a normally occurring metabolite of CCl$_4$ degradation in denitrifying environments. The field experiment was designed to adjust aquifer pH to create a niche for a *Pseudomonas* denitrifying bacterium that degrades CCl$_4$ to carbon dioxide, chloride, and non-volatile organics. An injection well with a screened interval of ~18-20 m (~60-65 ft) below land surface (BLS) was used to deliver successive injections of a pH adjustment/nutrient solution and a bacterial culture suspension. Monitoring wells with similar screened intervals were placed in two downgradient arcs at distances of 1 and 2 m (3 and 6 ft) from the injection well and equipped with multi-level diffusion cell samplers (MLS, Margan, Ltd.). Hydraulic conductivities at the site ranged from 0.0001-0.0005 cm/sec (0.3-1.5 ft/day), reflecting the typical properties of a sandy water table aquifer. Bulk flow rates averaged 15 cm/day (0.5 ft/day) and were not appreciably altered in magnitude or direction during the course of the experiment.

The evaluation plan called for pre-treatment ground water and sediment sampling and analysis, followed by dissolved-phase process monitoring for 4 months, and finally post-treatment sampling and analysis. The basis for interpretation of the results resided in pre- and post-treatment sediment samples collected from locations above, in, and below the intended treatment zone (*i.e.*, the screened interval). These results were coupled with fine-scale (15-20 cm or 6-8 inch interval), multi-level ground-water samples collected from the monitoring wells during the course of the experiment. Breakthrough of the injected tracer, nutrient, and bacterial suspension solution provided evidence of hydraulic communication between the injection well and the monitoring wells along pathways through the treatment zone.

The average and standard deviation of dissolved CCl$_4$ concentrations in ground water in the treatment zone before treatment were 52.4 ± 3.0 µg/L (n=5). At an effective porosity of 0.2, a representative one-liter unit volume of aquifer would thus contain 10.5 µg CCl$_4$ in the dissolved phase. This contrasts with an average sediment-associated CCl$_4$ concentration of 24.9 ± 38.9 µg/kg (n=58), yielding 35 µg CCl$_4$ on the solid phase if a minimum observed bulk density of 1.54 g/cm^3 is assumed to apply throughout the test site. CCl$_4$ sorbed onto the aquifer solids thus comprised an average of 77% of the total contaminant mass within the unit volume. The pre-treatment sediment-associated CCl$_4$ measurements are depicted in Figure 1. They clearly show substantial scatter throughout the sampled interval and an apparent trend of increasing average concentration with depth. The latter point is supported by a two sample student's t-test, which shows a statistically significant difference between samples taken at 60-62.5 ft BLS and 62.5-65 ft BLS versus samples taken from deeper 2.5 ft

202

increments (α=0.05), eliminating the region above the treatment zone as a potential experimental control. No significant difference was noted between the treatment zone and the other intended control region at 65 to 70 ft BLS.

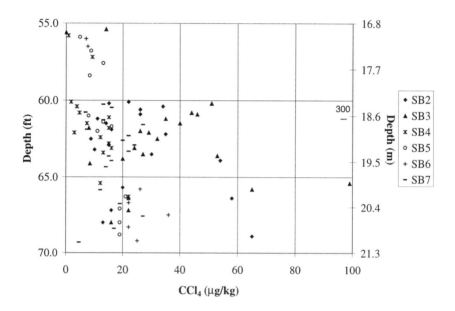

FIGURE 1. **Pre-treatment sediment-associated carbon tetrachloride measurements. Samples were taken from continuous cores collected adjacent to monitoring wells located 1 to 2 meters downgradient from the injection well.**

During treatment, dissolved-phase concentrations of CCl_4 declined to near detection limits, particularly in vertical intervals where high concentrations of the injected materials were observed, demonstrating good hydraulic communication with the injection well. Post-treatment sediment-associated CCl_4 results are depicted in Figure 2, and show generally lower concentrations with significantly less scatter than the measurements depicted in Figure 1. It is immediately obvious that the region below the treatment zone did not serve as a valid experimental control. At the conclusion of the experiment, the average sediment-associated CCl_4 concentration within the treatment zone had decreased to 9.9 ± 1.8 µg/kg (n=32). The greatly reduced standard deviation and smaller size of this second sample population is problematic, in that it violates the assumptions underlying many commonly used parametric statistical tests. A two-sample student's t-test assuming unequal variance can be used to test for a statistically significant decrease in CCl_4 concentrations within the treatment zone before and after treatment, which proved to be the case (α=0.05). However, this test is much more rigorous than the standard student's t-test that could have been applied if sample populations in the pre- and post- treatment datasets had been of equal size.

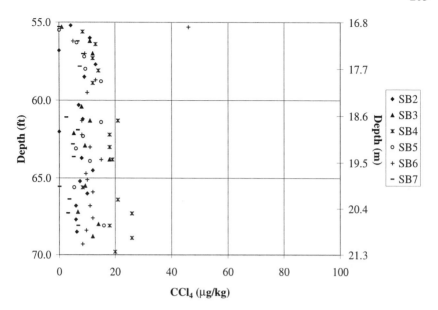

FIGURE 2. **Post-treatment sediment-associated carbon tetrachloride measurements. Samples were taken from continuous cores collected adjacent to monitoring wells located 1 to 2 meters downgradient from the injection well.**

The importance of using appreciable and comparable levels of sample replication to achieve sufficient precision in a technology evaluation becomes apparent in the results of this comparatively well-sampled experiment when a volumetrically averaged estimate of treatment efficiency is made. If error propagation is used to assign a standard deviation to an estimate of treatment efficiency with respect to sediment associated CCl_4, the removal efficiency is calculated to have been 60% ± 65%. If these parameters were assumed, within reason, to be invariant with decreasing size of comparable sample populations, the 95% confidence intervals assigned to the treatment efficiency estimate would be 60% ± 17% (n=58), 60% ± 23% (n=32), 60% ± 34% (n=14), or 60% ± 45% (n=8). Thus, in the face of natural and experimental variability, sample replication is a critical factor in the precision and reliability of the reported results.

Pilot Treatment Evaluation Example – Alternative Methods

Alternative methods of contaminant mass estimation were used in a recently conducted in-situ solubilization experiment by Jawitz *et al.* (*39*). A catastrophic glacial outwash deposit at Hill Air Force Base, Utah had become contaminated with a complex (>200 compounds) NAPL mixture derived from chemical disposal pit leachate. The field experiment evaluated the ability of a pump and treat system flushing a Winsor Type 1 surfactant/alcohol mixture to extract the highly hydrophobic NAPL as a single phase microemulsion (SPME). A 4.6 m by 2.8 m (15.1 ft by 9.2 ft) test cell was isolated from the aquifer by the use of interlocking steel sheet piles keyed into the underlying clay (~8 m or ~26 ft BLS). The cell was instrumented with four injection wells and three extraction wells located at opposite ends of the cell. A 3% polyoxyethylene(10) oleyl ether/2.5% 1-pentanol solution was selected as the surfactant/alcohol mixture to be flushed through the test cell, followed by a 3% surfactant solution and then unamended ground water to conclude the treatment phase of the experiment.

The volume of NAPL within the test cell before and after treatment was estimated by paired nonpartitioning (methanol) and partitioning (2,2-dimethyl-3-pentanol) tracer tests. The test cell was injected with tracer solution and flushed with unamended ground water until tracer concentrations declined below detection limits. Integration of tracer breakthrough curves at the extraction wells was used to estimate both the swept volume and NAPL saturation, which could be used to estimate the volume of NAPL within the test cell. The swept volume estimate was a particularly interesting feature of this experiment given the nearby, irregular boundaries that were a necessary part test cell construction. Large discrepancies between the assumed void volume and the swept volume could have indicated the existence of substantial 'dead zones' in the flow regime or the presence of regions of substantially lower hydraulic conductivity, aiding in the interpretation of point sample data.

Performance of the system was evaluated by several methods: through changes in the estimated volume of NAPL within the cell, changes in the sediment associated concentrations of 12 volatile and semi-volatile NAPL components, and integration of component breakthrough curves obtained from regular sampling of the produced fluids. A summary of selected data, including standard deviations, is included in Table II.

Table II. Winsor Type I Flushing Experiment – Sampling Results

Sampling Method/ Component	n	Average Initial Concentration	n	Average Final Concentration or Mass Removed	Average Removal Fraction[1]
Partitioning Tracer					
NAPL saturation	3	0.062 ± 0.008	3	0.017 ± 0.003	0.72 ± 0.06
Sediment Sampling (mg/kg sediment)					
m-xylene	83	0.12 ± 0.15	36	0.02 ± 0.04	0.87 ± 0.39
o-xylene	83	0.18 ± 0.31	36	0.06 ± 0.09	0.65 ± 0.79
1,2-dichlorobenzene	83	0.74 ± 1.99	36	0.03 ± 0.08	0.96 ± 0.16
1,3,5-trimethylbenzene	83	0.91 ± 1.23	36	0.11 ± 0.27	0.88 ± 0.34
Naphthalene	83	2.90 ± 2.49	36	0.25 ± 0.89	0.92 ± 0.31
n-decane	83	24.9 ± 21.2	36	1.10 ± 2.71	0.96 ± 0.12
n-undecane	83	61.2 ± 51.6	36	2.51 ± 5.87	0.96 ± 0.10
Breakthrough Integration (kg/kL swept volume)					
p-xylene	1[2]	0.08	1,3	0.05 ± 0.005	0.60 ± 0.057
1,2,4-trimethylbenzene	1[2]	0.25	1,3	0.19 ± 0.020	0.74 ± 0.081
n-decane	1[2]	0.27	1,3	0.19 ± 0.023	0.68 ± 0.086
n-undecane	1[2]	0.90	1,3	0.52 ± 0.056	0.57 ± 0.062
n-decane	83[3]	0.40 ± 0.34	1,3	0.19 ± 0.023	0.47 ± 0.082
n-undecane	83[3]	0.99 ± 0.84	1,3	0.52 ± 0.056	0.52 ± 0.099

SOURCE: Adapted from ref. 39.

[1] Standard deviations were propagated by $STD_r = [(Mean_r STD_i/Mean_i^2)^2 + (STD_f/Mean_i)^2]^{1/2}$

[2] Estimate based on component mass fraction in a NAPL sample and the swept volume and NAPL volume indicated by the initial partitioning tracer test. Standard deviations are for illustration only.

[3] Estimate based on sediment sampling results, swept volumes, a measured effective porosity of 0.14, and an assumed mineral density of 2.64 g/cm³. Table entries are for illustration only.

The pre-treatment volume-weighted average NAPL saturation in the test cell was 0.061. If the data obtained from the three extraction wells are instead treated as three observations of the same sample population, then the average and standard deviation of NAPL saturation in the test cell was 0.062 ± 0.008. This measurement was repeated after SPME treatment to provide an estimate of the reduction in NAPL saturation. It is important to note that the use of the tracer test allowed the analysis to be performed upon a very large sample volume and thus it presumably provides a more accurate estimate of NAPL saturation than a similar population of smaller samples. However, the greatly reduced sample population resulted in a less reliable estimate of the precision of the pre- and post-treatment determinations.

Estimates of the removed mass of NAPL components obtained by sediment sampling and analysis were subject to effects similar to those discussed in the previous example. Of the 12 NAPL components selected as indicator compounds for point sample analysis, 7 were present within the pre-operational test cell at levels (>0.1 mg/kg) suitable for interpretation. These results were obtained by using a much smaller sample size but more numerous sample population (n=83) and reveal the existence of substantial variability in contaminant distributions within the test cell. Post-operational results (n=36) display similar levels of relative variability, yet when evaluated with a two-sample student's t-test assuming unequal variance they indicate a statistically significant (α=0.05) reduction in sediment associated concentrations of the indicator compounds.

Jawitz *et al.* (*39*) also used integration of dissolved-phase component breakthrough curves, in combination with an analysis of NAPL composition and a NAPL volume estimate obtained from the partitioning tracer test, to calculate the removal fractions of several components. It is important to note that these removal fractions were based upon a single observation, the integrated total, of the mass removed by the experiment and thus were reported by the authors as estimates, lacking a basis for the calculation of a mean, standard deviation, or confidence interval. By normalizing the masses withdrawn at the three extraction wells to the swept volumes observed in the two partitioning tracer experiments, natural variability can begin to be estimated. However, the standard deviations reported in Table II for components analyzed by this method are incorrect. While they capture some of the natural and experimental variability in the experiment, the estimates of the initial masses in the test cell fail to do so. Thus the standard deviations reported for the removal fractions determined by this method underreport the true magnitude of variability in the experiment. Initial mass estimates that incorporate natural variability can be obtained by using the sediment sampling data, but in this case the calculation relies upon the use of additional single valued estimates of test cell properties. Furthermore, after this degree of manipulation and considering the disparate sources of the figures involved, the statistical comparability of the data is questionable at best and confidence intervals are essentially incalculable.

As this example illustrates, there are substantial differences in the characteristics of various mass estimation methods. In addition to ensuring that the methods used to analyze the 'before and after' states of an experiment are comparable, the investigator will often need to make complex design decisions that acknowledge tradeoffs between achievable levels of accuracy and precision given the resources available to an experiment. The accuracy and precision of a particular technique may be superior, but the expense and resultant lack of sample replication may make the confidence intervals associated with the result poorer than those that could be obtained with a less expensive technique and a larger sample population.

The use of a partitioning tracer to estimate NAPL volume in the test cell illustrates an important aspect of the use of surrogate measures of contamination in technology evaluations. The authors examined the character of the remaining NAPL and noted that it was neither water soluble nor extractable in hexane, calling it a 'pitch' fraction. The type of regulatory standard governing the remediation of a site will govern whether overall contamination (*e.g.*, TPH) must be reduced below numerical limits, contamination by chemicals of concern (*e.g.*, BTEX, chlorinated solvents) must be reduced below numerical limits, and/or risk of human exposure and health effects must be reduced below a specified probability level. In each of these cases, performance of the surrogate indicator must be carefully related to the critical regulatory and/or design goals of the candidate technology. In this case, the probability of human exposure to the remaining NAPL fraction was reduced far beyond the level indicated by the 72% reduction in NAPL volume achieved by SPME flushing.

Future Directions

Innovative technology field testing has advanced to the quantitative stage given the level of funding available over the past decade. Past evaluations provide valuable design datasets with which more robust experimental designs can be developed. The use of surrogate field analytical techniques will provide the levels of sample density needed to support more rigorous difference testing. Future evaluations should provide more reliable cost estimates for full-scale operational remedial actions which can be used in risk-cost based cleanup decision making. The public acceptance of inevitable trade-offs can then be founded on a more technically defensible basis.

Literature Cited

1. Martin-Hayden, J. M.; Robbins, G. A. *Ground Water* **1997**, *35* (2), 339-346.
2. Barcelona, M. J. In *Principles of Environmental Sampling*; Keith, L. H., Ed.; ACS Professional Reference Book; American Chemical Society: Washington, D.C., 1996, 2nd edition; 41-62.
3. Barcelona, M. J. *Site Characterization: What Should We Measure, Where, (When?) and Why*; EPA 540/R-94/515; USEPA-RS Kerr Laboratory, U.S. EPA: Ada, OK, 1994.
4. Robbins, G. A. *Ground Water* **1989**, *27* (2), 155-162.
5. Robbins, G. A.; Martin-Hayden, J. M. *J.Contam. Hydrol.* **1991**a, *8* (3-4), 203-224.
6. Martin-Hayden, J. M.; Robbins, G. A. *J.Contam. Hydrol.* **1991**b, *8* (3-4), 225-241.
7. Gelhar, L. W.; Welty, C.; Rehfeldt, K. R. *Water Resour. Res.* **1992**, *28* (7), 1955-1974.
8. Rehfeldt, K. R. et al. *Water Resour. Res.* **1992**, *28* (12), 3309-3324.
9. Sudicky, E. A. *Water Resour. Res.* **1986**, *22* (13), 2069-2082.
10. Bjerg, P. L.; Christensen, T. H. *J. Hydrol.* **1992**, *131*, 133-149.
11. Ronen, D.; Margaritz, M.; Levy, I. *Ground Water Monit. R.* **1987**, *7* (4), 69-74.
12. Mercer, J.; Cohen, R. M. *DNAPL Site Evaluation*; C.K. Smoley CRC Press: Boca Raton, FL, 1993.
13. Boulding, J. R. *Site Characterization for Subsurface Remediation*; EPA/625/4-91-026; Center for Environmental Research Information, Office of Research and Development, U.S. EPA: Cincinnati, OH, 1991.
14. Anderson, W. C. *Remediation technologies Screening Matrix and Reference Guide*; American Academy of Environmental Engineers: Annapolis, MD, 1993.
15. U.S. EPA *International Evaluation of In Situ Biorestoration of Contaminated Soil and Groundwater*; EPA/540/2-90/012; Office of Emergency and Remedial Response and Office of Research and Development, U.S. EPA: Washington, D.C., 1990.
16. Wiedemeier, T. H.; Wilson, J. T.; Kampbell, D. H.; Kerr, R. S.; Miller, R. N.; Hansen, J. E. *Technical Protocol for Implementing Intrinsic Remediation with Long-Term Monitoring for Natural Attenuation of Fuel Contamination Dissolved in Groundwater*; Air Force Center for Environmental Excellence: Brooks AFB, San Antonio, TX, 1995.

17. AATDF (Advanced Applied Technology Demonstration Facility Program) *AATDF Technology Practices Manual for Surfactants and Cosolvents*; AATDF Report TR-97-2; Rice University Energy and Environmental Systems Institute: Houston, TX, 1997.

18. Martin-Hayden, J.M.; Robbins, G.A., *Ground Water* **1997**, *35* (2), 339-346.

19. Hewitt, A. D. *Losses of TCE from Soil During Sample Collection, Storage and Lab Handling*; SR94-8; Cold Regions Environmental Laboratory, U.S. Army Core of Engineers: Hanover, New Hampshire, 1994.

20. Hewitt, A. D.; Miyares, P. H.; Leggett, D. C.; Jenkins, T. F. *Environ. Sci. Technol.* **1992**, *26* (10), 1932-1938.

21. Siegrist, R. L *J. Hazard. Materials* **1992**, *29* (**1**), 3-15.

22. Siegrist, R. L.; Jenssen, P. D. *Environ. Sci. Technol.* **1990**, *24* (9), 1387-1392.

23. Smith, J. S. In *Principles of Environmental Sampling*; Keith, L. H., Ed.; American Chemical Society: Washington, D.C., 1996, 2nd edition; 693-704.

24. Caughey, M. E.; Barcelona, M. J.; Powell, R. M.; Cahill, R. A.; Gron, C.; Lawrenz, D.; Meschi, L. *Environmental Geology* **1995**, *26* (4), 211-219.

25. Barcelona, M. J.; Lu, J.; Tomczak, D. M. *Ground Water Monitoring and Remediation* **1995**, *15* (2), 114-124.

26. Sale, T. C.; Applegate, D. H. *Wat. Env. Res.* **1996**, *68* (7), 1116-1122.

27. Sale, T. C.; Applegate, D. H. *Ground Water* Submitted for publication.

28. U.S. EPA *Superfund Innovative Technology Evaluation Program and the Inventory of Treatability Study Vendors*; EPA/540/2-90/003b; Office of Solid Waste and Emergency Response, U.S. EPA: Washington, D.C., 1990.

29. SERDP (Strategic Environmental Remediation Defense Program); TRW, Inc.; Praxis Environmental Technologies, Inc. *Guidelines for Quality Technology Demonstrations*; MDA970-89-C-0019; SERDP: Washington, D.C., 1996.

30. U.S. Air Force *Remedial Technology Design, Performance and Cost Study*; U.S. Air Force Center for Environmental Excellence: Brooks AFB, San Antonio, TX, 1992.

31. Boulding, J. R.; Barcelona, M. J. *In-Situ Characterization for Subsurface Remediation*; EPA 625/4-91/026; Center for Environmental Research Information, U.S. EPA: Cincinatti, OH, 1989.

32. Havlicek, L. L.; Crain, J. D. *Practical Statistics for the Physical Sciences*; ACS Professional Reference Book; American Chemical Society: Washington, D.C., 1988.

33. Pitard, F. F. *Pierre Gy's Sampling Theory and Practice*; CRC Press: Boca Raton, FL, 1994; 2nd edition.

34. Starks, T. H.; Brown, K. W.; Fisher, N. J. In *Quality Control in Remedial Site Investigation*; Perket, C. L., Ed.; ASTM, STP #925; American Society for Testing and Materials: Philadelphia, PA, 1986; 57-66.

35. Gilbert, R. O.; Simpson, J. C. *Env. Monit. Asses.* **1985**, *5*, 113-135.

36. Gilbert, R. O. *Statistical Methods for Environmental Pollution Monitoring*; van Nostrand Reinhold Co.: New York, NY, 1987.

37. Flatman, G. T.; Yfantis, A. A. In *Principles of Environmental Sampling*; Keith, L. H., Ed.; ACS Professional Reference Book; American Chemical Society: Washington, D.C., 1996, 2nd edition; 779-801.

38. Dybas, M. J. *Evaluation of Bioaugmentation to Remediate an Aquifer Contaminated with Carbon Tetrachloride*; Battelle Symposium on In-Situ and Onsite Bioremediation **1997**, *4*, 507-512.

39. Jawitz, J. W.; Annable, M. D.; Rao, P. S. C.; Rhue, R. D. *Environ. Sci. Technol.* **1998**, *32* (4), 523-530.

Chapter 15

The Partitioning Tracer Method for In-Situ Detection and Quantification of Immiscible Liquids in the Subsurface

Mark L. Brusseau[1,2], N. T. Nelson[2], and R. Brent Cain[2]

[1]Departments of Soil, Water, and Environmental Science and [2]Hydrology and Water Resources, University of Arizona, Tucson, AZ 85721

Conducting risk assessments and developing remediation programs for contaminated sites requires knowledge of the occurrence and distribution of immiscible liquids in the subsurface. Current "point-sampling" characterization methods, such as analysis of soil gas, core sampling, cone penetrometer testing, and monitor-well sampling, are often not capable of providing the amount of data required for effective +risk assessment or remediation system design, without a cost-prohibitive number of samples. The purpose of this work is to discuss an alternative method, the partitioning tracer test, for measuring the quantity and distribution of immiscible liquid in the subsurface. The basis for this method will be briefly reviewed, followed by a discussion of its implementation. Practical considerations as well as limitations of the method will also be discussed. Application of the method will be illustrated for a site contaminated with a multiple-component immiscible liquid.

The contamination of soil and groundwater by hazardous organic chemicals and the associated risks to humans and the environment have become issues of great interest and importance. Much of the research motivated by subsurface contamination has been focused on dissolved constituents in aqueous systems. However, nonaqueous phase liquids (NAPLs) are also of great concern and may occur at many contaminated sites. Spills and leaks of petroleum-based liquids such as gasoline, fuel oil, and jet fuel are a major source of NAPL. Coal-tar and creosote, associated with manufactured-gas plants and wood-treatment facilities, are another major source of NAPL. Chlorinated solvents used in industrial and commercial activities are a third major source of NAPL. It is now widely accepted that NAPL saturation in the subsurface can be a long-term source of both vapor-phase and groundwater contamination.

According to a recent report released by the National Research Council, the presence of NAPL is the single most important factor limiting site clean-up. In fact, groundwater restoration to health-based goals is considered infeasible with existing technologies (e.g., pump and treat) at most sites contaminated by dense NAPL (1). Thus, new technologies continue to be developed in an attempt to address this problem. The development and application of new remediation technologies requires an understanding of the mechanisms that control the transport and fate of NAPLs, as well as the nature and distribution of the NAPL in the subsurface. Knowledge of NAPL occurrence and distribution is also critical for the planning and implementation of "source control" strategies. Additionally, accurate risk assessments can not be made for a contaminated site without knowing if NAPL is present at the site.

Current characterization methods, such as analysis of soil gas, core sampling, cone penetrometer testing, and monitor-well sampling, are often not capable of providing the level of data required for effective risk assessment or remediation system design. A major weakness of these methods is that they provide data at discrete points, such that the probability of sampling a zone of localized NAPL is quite small. In addition, because the distribution of NAPL is complex and highly variable, it is difficult to use data collected at discrete points to make accurate predictions about the overall contaminant distribution without a cost-prohibitive amount of sampling.

The purpose of this work is to discuss an alternative method, the partitioning tracer test, for measuring the quantity and distribution of immiscible liquid in the subsurface. The basis for this method will be briefly reviewed, followed by a discussion of its implementation. Practical considerations as well as limitations of the method will also be discussed. Application of the method will be illustrated for a site contaminated with a multiple-component immiscible liquid.

Background

The partitioning tracer test is based on conducting a tracer test in the targeted zone of the subsurface. The potential presence of NAPL is evaluated by comparing the transport of one or more partitioning tracers to that of a conservative (nonpartitioning) tracer. Organic-fluid phases can reversibly retain the partitioning tracers, which retards their transport with respect to that of the conservative tracer. Thus, retardation of the partitioning tracers indicates the possible existence of NAPL saturation within the zone through which the tracer solution moved (swept zone). When other potential sources of tracer retention, such as sorption by the aquifer solids, are negligible or have been accounted for, the magnitude of the measured retardation of the partitioning tracer can be translated to a quantitative measure of the NAPL saturation.

The theoretical basis for the partitioning tracer method was established in the chromatography field by Martin and Synge (2), who developed the liquid-liquid chromatography technique. In this technique, compounds dissolved in a mobile liquid are separated as they pass through a column containing an immobile, immiscible liquid, for which the magnitude of retention of each compound is different. The use of partitioning tracers to measure NAPL saturation in the subsurface was pioneered by the petroleum industry as a means to determine residual oil saturation in oil fields (3,4). Since that time, upwards of 300 partitioning tracer tests have been conducted, as reviewed by Tang (5).

In the environmental field, the retention of dissolved solutes by immobile, immiscible-liquid phases and the resultant impact on solute (tracer) transport was initially described by Bouchard et al. (6) and Brusseau (7,8). Using miscible-displacement experiments, Bouchard et al. (6) demonstrated that the retardation of toluene and three other gasoline constituents was greater when the aquifer material contained a residual saturation of aviation gas. Brusseau (7) reported similar results for the transport of naphthalene in a column containing aquifer material contaminated with a residual saturation of tetrachloroethene. The influence of the magnitude of NAPL saturation and the magnitude of the liquid-liquid partition coefficient on the retardation of a partitioning tracer was illustrated with numerical-model simulations reported by Brusseau (8). For example, the influence of 0, 1, and 10% saturation of a hypothetical NAPL on the transport of a partitioning solute is illustrated in Figure 1.

The use of the partitioning tracer method for measuring NAPL saturation in environmental systems was introduced and tested in the laboratory by Jin et al. (9) and Wilson and Mackay (10). Jin et al. (9) demonstrated that partitioning tracer tests using two alcohols provided good estimates of the amount of liquid tetrachloroethene in columns packed with sand. Wilson and Mackay (10) used sulfur hexafluoride as a partitioning tracer to successfully predict the trichloroethene saturation in columns packed with sand.

The application of the partitioning tracer method at the field scale has recently been demonstrated at several contaminated sites. As discussed in several chapters within this book, pilot-scale field experiments were conducted at Hill Air Force Base in Utah to test the use of several alcohol tracers for measuring the amount of petroleum-based NAPL contained within 3 m by 5 m fully enclosed cells (11). For each cell, partitioning tracer tests were conducted before and after an experiment designed to test the performance of a specific remediation technology. The results obtained from the partitioning tracer tests were used to help evaluate the mass (volume) of NAPL removed by the remediation technology. Results indicate that the NAPL saturation values obtained from the partitioning tracer tests are in good agreement with values obtained from core analysis (e.g., 12, 13).

In the applications discussed above, partitioning tracers were used to measure quantities of known NAPL contamination. However, it is feasible that partitioning tracers may also be useful as "detectors" of NAPL saturation. Such a use was demonstrated recently by Nelson and Brusseau (14) for a chlorinated-solvent contaminated aquifer at a Superfund site in Tucson, AZ. The results of the partitioning tracer tests indicated retardation of the partitioning tracer (SF_6) relative to the nonreactive tracer (bromide) at certain monitoring locations (see Figure 2 for an example). At other locations, the SF_6 breakthrough curves were coincident with those of bromide, indicating no sorption of SF_6 by the solid phase (which was consistent with results of laboratory experiments). The results of the partitioning

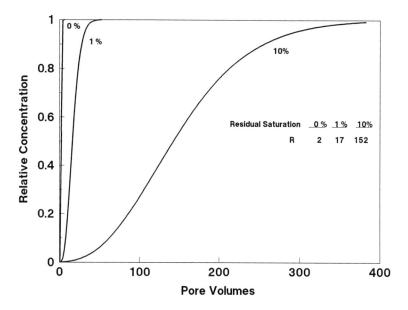

Figure 1.   The impact of immobile, immiscible-liquid saturation on the transport and retention of a solute that partitions between water and the immiscible liquid; from Brusseau (8).

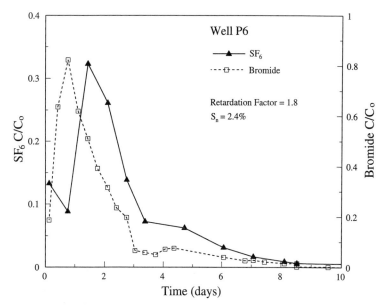

Figure 2.   Breakthrough curves measured for transport of bromide (nonreactive) and SF$_6$ (partitioning) tracers in an aquifer located below a source zone of a chlorinated-solvent contaminated Superfund site; from Nelson and Brusseau (14).

tracer tests indicated the probable presence of NAPL saturation within the swept zone of the tests. These results were in accordance with other lines of evidence, such as the removal of large volumes of solvent during the operation of a soil venting system in the vadose zone directly above the location of the tracer tests.

Implementation

Interpretation of Partitioning Tracer Tests. Partitioning tracer tests can be used to detect the presence of NAPL saturation, as discussed above. By sampling a much larger volume of the subsurface compared to that measured with cores or monitoring wells, the partitioning tracer method has a much greater chance of detecting NAPL saturation. This increased effectiveness for detection of NAPL saturation is a major advantage of the method. The sensitivity of the partitioning tracer method for NAPL detection is a function of the area of influence of the tracer test (swept volume), the amount of NAPL in that swept volume, the NAPL-water partition coefficient of the tracer, and constraining factors (rate-limited mass transfer, heterogeneity, mass loss). If the ratio of NAPL saturation to swept volume is too small for a given set of tracers, the influence of the NAPL saturation on tracer retardation could be lost within the normal uncertainty associated with field data.

Partitioning tracer tests can also be used to measure or quantify the amount of NAPL saturation located within a target zone. The procedure for estimating S_n, NAPL saturation, involves calculation of a retardation factor, R, for the partitioning tracer. This calculation can be done in a number of ways, as will be discussed in the following paragraph. With knowledge of the NAPL-water partition coefficient (K_n), soil-water partition coefficient (K_d), bulk density of the porous medium (ρ_b), volumetric water content (θ_w), and the measured R, S_n can be calculated using the following definition of the retardation factor (e.g., Brusseau, 8):

$$R = 1 + \frac{\rho_b K_d}{\theta_w} + \frac{S_n K_n}{(1-S_n)} \tag{1}$$

When there is negligible sorption of the tracer by the soil, $K_d = 0$ and:

$$R = 1 + \frac{S_n K_n}{(1-S_n)} \tag{2}$$

An alternative approach based on travel times can also be used (e.g., Jin et al., 9). The two are equivalent through the equality: $R = t_p/t_t$, where t_p and t_t are the mean travel times for the partitioning and nonreactive tracers, respectively.

The typical manner of analysis for a partitioning tracer test involves a comparative moment analysis of the partitioning and conservative tracers. In this approach, the retardation factor is defined as the quotient of the mean travel time of the partitioning tracer and the mean travel time of the conservative tracer. However, travel times or retardation factors can also be obtained by other methods. For example, with the "landmark" method (5), calculation of travel times is based on comparing specific points of the respective breakthrough curves, such as times of first arrival, times of peak arrival, or times of selected magnitudes of mass recovery. Mathematical models can also be used to obtain retardation factors. The advantage of the moment method is that it may be less sensitive to nonideality factors (rate-limited mass transfer, heterogeneity effects) than the landmark method, and it is not dependent upon a suite of assumptions as are modeling-based analyses. However, the moment method is susceptible to error, especially when breakthrough curves are incomplete, when mass transfer constraints are extreme and, in some cases, when mass loss occurs.

The measured S_n values may often be underestimates of the true values because of factors such as bypass flow (water flows around a NAPL zone due to reduced relative permeability), rate-limited mass transfer, and mass loss. Thus, NAPL saturation measurements obtained with the partitioning tracer method should, at least initially, be considered as underestimates of actual values. In addition, when interpreting the results of a partitioning tracer test, it is important to realize that the S_n values obtained from the test are "global" values that are averaged across the measured domain (i.e., swept zone). Retardation of the partitioning tracer indicates (possible) saturation somewhere within the swept volume, but its exact location is not revealed.

Tracer-Test Methods. The partitioning tracer test is based on methods that are essentially identical to those used for nonreactive tracer tests. Thus, the large body of knowledge in Hydrology and Petroleum Engineering regarding tracer tests is applicable. Tracer tests can be implemented in a number of ways, based primarily on the number and configuration of the wells. The simplest configuration involves a single well, wherein the tracer solution is first injected into and then extracted from the aquifer (push-pull test). Water samples for analysis of tracer concentrations are collected at the well. Because of the reversible retention of the partitioning tracers used in a partitioning tracer test, a special approach is used for single-well push-pull tests. This method, described by Deans (4), involves the use of a partitioning tracer that hydrolyzes to a compound that has a different magnitude of NAPL partitioning. After the tracer is pumped into the aquifer, the solution is allowed to remain in the system for a period of time to allow transformation (the so-called "shut-in" period). The solution is then pumped back to the well, and samples are analyzed for concentrations of both the parent and product. This method allows the two tracers to start at the same position within the aquifer, thus providing an opportunity for separation to occur during transport.

A variation of the single well system involves the use of multiple well screens, i.e., so-called vertical circulation wells. The placement of two or more screened intervals within a single well allows one interval to be used for injection and one interval for extraction. A vertical dipole is created with this configuration, which produces vertically oriented flow. This system can be used to conduct single-well partitioning tracer tests, as demonstrated by Chen and Knox (15) for a sand-tank system and by Cain et al. (13) at the Hill AFB field site.

Another well configuration involves a single extraction well and a non-pumping injection well, where the passive well is used to introduce the tracer solution into the flow field created by the extraction well, which is used for sampling. Another variation involves a single pumping injection well and one or more monitoring wells surrounding the injection well. Given the use of separate injection and extraction points, the use of transforming tracers is not required for either of these, or the vertical circulation well, configurations.

The fourth basic design for tracer tests involves a combination of pumping injection and extraction wells. The use of this configuration, referred to as an interwell tracer test, was first proposed for the partitioning tracer method by Cooke (3), and is the most common method for environmental applications. The utility and effectiveness of different well configurations for nonreactive tracer tests was evaluated by Gelhar et al. (16), who concluded that the dual injection-extraction well configuration provided the most reliable data. Of course, the method used at a given site will depend on the objectives and limitations associated with that site.

The use of monitoring wells is rarely addressed in the petroleum applications of partitioning tracer tests. However, monitoring wells are a common feature at environmental field sites, and their use in partitioning tracer tests can provide great benefits. For example, as mentioned previously, there is a limit to the sensitivity of the partitioning tracer method. The sensitivity depends in part on the ratio of NAPL mass to swept volume. The swept volume of an extraction well can be very large. Thus, a small, localized mass of NAPL could have a very small impact on retardation as measured at an extraction well. This small signal could be lost in the noise (i.e., data scatter) typically associated with field experiments. Conversely, the swept volume is relatively small for monitoring wells, especially those located close to the injection zone. Thus, a relatively small, localized mass of NAPL may be more easily "measured". However, an extraction well "measures" a much larger area, by virtue of its larger swept zone which, as previously discussed, is an advantage when using the partitioning tracer method to search for NAPL. Thus, it is useful, if possible, to use both monitoring wells and extraction wells for sampling.

As discussed previously, the saturation values obtained from a partitioning tracer test are averaged over the entire swept zone between the injection and sampling points. Monitoring wells, given their smaller swept volumes, can be used to focus on specific sections of the total swept zone associated with the injection/extraction wells. This provides an opportunity to examine spatial variability of NAPL saturation, albeit at a relatively large scale. In addition, as shown by Nelson and Brusseau (14), multilevel sampling devices can be used to examine the vertical distribution of NAPL occurrence. Thus, a three-dimensional distribution of NAPL saturation can be developed by using several monitoring wells, each equipped with a multilevel sampling device.

The tracers to be used in the test are selected to meet several criteria. The magnitude of retardation of the partitioning tracers is a key factor. The retardation should be large enough ($>$ ~1.2) to be measurable with some level of certainty, but should be small enough such that the length of the test remains practical. The length of the test will of course depend on the size of the target domain and the planned flow rates, as well as the magnitude of retardation. The magnitude of retardation is a function of the NAPL saturation and the partition coefficient. If an approximate range of NAPL saturations expected to be encountered is known, the partitioning tracers can be selected according to their partition coefficients to provide retardation factors within the desired range. If the NAPL saturation is unknown, it may be useful to use multiple partitioning tracers with different partition coefficients, thus providing a range of potential retardation.

When multiple partitioning tracers are used, the travel-time analysis can be conducted using the various pairs of partitioning tracers, as well as the partitioning-nonpartitioning pairs. This approach can in some cases increase the robustness of the results. The partitioning tracers, as well as the conservative tracers, should also be non-toxic at the concentrations employed, resistent to mass loss mechanisms, and relatively inexpensive.

Constraints

Difficulties in analyzing the transport behavior of partitioning tracers may arise due to nonideal transport caused by factors such as subsurface heterogeneity (e.g., spatially variable hydraulic conductivity), variable distribution of NAPL, structured soil, rate-limited mass transfer, and tracer mass loss. For example, extreme rate-limited mass transfer can cause extensive concentration tailing, which in turn can prevent the accurate calculation of moments (e.g., 8). The impact of these factors on the partitioning-tracer method have not yet been examined in detail.

Mass transfer between the NAPL and aqueous phases is a controlling factor for the partitioning tracer method. The rate of mass transfer and its effect on dissolution of trapped NAPL has been investigated as a function of aqueous phase velocity, fluid saturations, and porous media characteristics (e.g., 17-20). In contrast to the effect of mass transfer on NAPL dissolution, very little research has been conducted regarding the effect of mass transfer on the retention and transport of dissolved solutes (e.g., tracers) in systems containing immiscible liquids.

Brusseau (8) quantitatively investigated the effect of mass transfer between water and immobile immiscible organic liquids on the retention and transport of organic solutes. A general, one-dimensional mathematical model was developed that allows for rate-limited uptake/release of solute by an immobile organic phase. The model also accounts for heterogeneous or structured porous media and rate-limited sorption/desorption of solute by the porous-media solids. The work presents a specific model for a case where the mass transfer of the solute across the liquid-water interface and within the immiscible liquid is rate limited. It was found that liquid-liquid mass transfer may be rate limited under some conditions, especially for induced-gradient conditions such as those reflective of pump-and-treat groundwater remediation systems and some tracer experiments.

The degree to which liquid-liquid mass transfer of a partitioning tracer will be rate limited during a tracer test will depend on the relative magnitudes of the characteristic times of mass transfer and advective transport. The latter is represented by the hydrodynamic residence time of the tracer (equivalent to t_p), which is determined by the hydraulic residence time (flow rate and size of swept volume) and the magnitude of partitioning. For a given tracer and well-field system (e.g., fixed magnitude of partitioning and swept volume), the flow rate is the "design" parameter influencing the degree to which mass transfer may be rate limited. Thus, flow rates can be selected to minimize the impact of rate-limited mass transfer. However, it is critical to note that it is residence time, not flow rate, that actually determines the degree to which tracer transport may be influenced by rate-limited mass transfer. Flow rates equivalent to hydraulic residence times of about one pore volume per day were used in the Hill AFB tracer tests and in the partitioning tracer test reported by Nelson and Brusseau (14).

The liquid-liquid partitioning process is generally considered to be a linear, reversible process at the microscopic scale. However, there are several processes that can cause the partitioning to appear to be non-reversible at the field scale. Consider, for example, a NAPL body or zone that is relatively thick in the dimension normal to water flow. When a partitioning tracer first contacts the NAPL zone, there is a concentration gradient driving the tracer into the NAPL. The concentration gradient reverses when the tracer pulse is followed by tracer-free water, which causes the tracer to transfer back to the advecting water. However, if the size of the pulse was insufficient to allow the tracer to fully saturate the NAPL zone prior to the elution step, an inward concentration gradient will still exist in the interior of the NAPL zone. This could significantly delay the return of some of the tracer mass to the advecting water. Depending on the time scale of the experiment and the nature of the NAPL, this severely rate-limited mass-transfer behavior could cause reduced mass recovery and, thus, the appearance that partitioning was non-reversible.

As noted above, liquid-liquid partitioning is usually assumed to be linear. This will generally be true for many of the systems encountered in environmental applications. However, it is possible that partitioning may be "nonlinear" in some cases. For example, a partitioning tracer could possibly exhibit nonideal behavior in the NAPL phase, depending on its composition. In such cases, the magnitude of partitioning could be dependent upon the tracer concentration (e.g., concentration-dependent activity coefficient) or the composition of the NAPL, which may be spatially or temporally variable.

Recovery of the tracer can be influenced by hydraulic-related factors, such as failure to capture all injection-well flow lines. While the impact of hydraulic factors on tracer recovery should be considered, one purpose of using a nonreactive tracer is to attempt to account for such factors. Tracer recovery can also be influenced by true mass loss processes, such as biodegradation and abiotic transformations. Mass loss of the tracer can influence the measurement and calculation of NAPL saturation. The degree of impact will depend on the nature of the mass loss.

The impact of mass loss on performance of the partitioning tracer test is due to the influence of mass loss on the travel time (first temporal moment) of the solute. When the mass loss is nonlinear, the measured travel time may differ from what it would have been without mass loss. An example of such behavior was illustrated by

Brusseau et al. (21), who showed that the travel times measured for tracers undergoing nonlinear biodegradation were smaller than those measured for a nonbiodegradable tracer under identical conditions. As discussed above, NAPL saturation values are obtained by comparing the travel times measured for the partitioning and conservative tracers. If either of the tracers experience a nonlinear mass loss, the measured travel times may be incorrect, which would result in an incorrect calculation of NAPL saturation. If the partitioning tracer is preferentially degraded, S_n would be underestimated, whereas S_n would be overestimated if the nonreactive tracer was preferentially degraded.

Under many conditions, travel time is not significantly affected when mass loss is linear and uniform, i.e., when the proportional mass loss is constant at all times and for all locations (e.g., 22). Hence, the results of a partitioning tracer test may not be significantly influenced by mass loss if the mass loss mechanism is linear. It is important to evaluate the potential causes of mass loss, if such is observed, for each specific case.

The performance of a partitioning tracer test can also be influenced by heterogeneity related factors. For example, groundwater may flow around (bypass) zones of high NAPL saturation due to the reduced relative permeabilities associated with these zones. Such bypass flow may limit the contact of the tracer solution to the periphery of the NAPL zone, thereby resulting in an underestimation of NAPL saturation. Similar phenomena may occur for systems containing NAPL "pools". Media heterogeneity (e.g., permeability heterogeneity) may also constrain the effectiveness of partitioning tracer tests. For example, NAPL saturation present in low permeability zones within or adjacent to high permeability zones may be undermeasured due to the preferential flow that occurs in such systems.

In the discussions above, K_n was treated as temporally and spatially constant. This may not be the case, however, for some situations. The magnitude of K_n is determined by the physicochemical properties of the tracer and the immiscible liquid. Thus, for a given tracer, K_n is a function of the NAPL composition. The composition of a NAPL may vary as a function of space and time due to a number of factors. Hence, K_n values may be spatially and temporally variable. This should be considered in the application of a partitioning tracer test.

Additional Partitioning Tracer Test Methods

Gas-Phase Partitioning Tracer Tests. The discussion presented above focused on measuring bulk NAPL saturation in water-saturated zones using partitioning tracer tests based on water flow. A similar approach could also be used to measure NAPL saturation in the vadose zone. However, given the relative rates of water and gas flow, a more efficient approach would be to conduct gas-phase partitioning tracer tests. The major difference between gas and water tracer tests is the selection of the tracers to reflect the gas phase as the mobile fluid. The use of gas-phase partitioning tracer tests to determine residual oil saturation in gas-saturated petroleum reservoirs was introduced by Tang and Harker (23,24). More recently, the use of gas-phase partitioning tracer tests to measure NAPL saturation in the vadose zones of contaminated sites has been examined with laboratory (25) and field (26,27) experiments.

For example, we have conducted gas-phase tracer tests in the vadose zone of a fuel-contaminated site in Tucson, AZ (27). The transport of the NAPL-partitioning tracer was retarded compared to that of the conservative tracer (see Figure 3). From the magnitude of retardation, a value of approximately 0.3% was estimated for S_n. This level of saturation is consistent with the performance of a soil venting system that is operating at the site. For example, approximately 200,000 kg of NAPL was removed from the subsurface prior to the tracer test. Distributing this mass throughout the zone of influence of the soil venting system results in an effective "initial" saturation of about 2.5% (including the NAPL measured with the tracer test). This magnitude of saturation is within the range of values reported in the literature for field systems.

The partitioning tracer concept can be used for applications other than measuring bulk-phase NAPL saturation. For example, as discussed by Brusseau et al. (28), gas-phase partitioning tracer tests can be used to measure water content in the vadose zone. Characterizing soil-water content is critically important to many activities, including those associated with agriculture, forestry, hydrology, and engineering. The majority of the methods currently used for measuring water content, such as gravimetric assay of core samples, neutron probes, and tensiometers, provide "point values" of water content (small sample volumes). While this is an advantage for obtaining accurate information at small scales, it is a disadvantage for determining water contents for larger (field) scales. The partitioning tracer method can be used to efficiently measure water contents for larger scales.

The application of gas-phase water-partitioning tracer tests has been demonstrated at the laboratory (28) and field (26,29) scales. For example, Brusseau et al. (28) used helium and CO_2 as the conservative and water-partitioning tracers, respectively, for laboratory experiments using columns packed with sandy porous media. The transport of CO_2 was retarded compared to that of helium (see Figure 4). The water content determined from the retardation of CO_2 was 0.14, which is close to the gravimetrically measured value of 0.16.

Interface Partitioning Tracer Tests The NAPL-water, NAPL-air, and air-water interfacial areas are of great significance for contaminant transport. The interfacial areas are, in part, reflections of the pore-scale distribution of the fluids in the porous medium. As such, knowledge of the interfacial areas could provide insight to the movement and redistribution behavior of the fluids in the system. In addition, mass transfer across an interface is a function of the interfacial area. Thus, knowledge of the interfacial areas would provide greater understanding of mass-transfer phenomena. It is possible that the magnitudes of interfacial areas may be evaluated by use of an interface partitioning tracer test. The application of interface partitioning tracers has recently been discussed for water flow (30) and gas flow (28) systems.

It is well known that many organic molecules (surfactants, long-chain alcohols) will tend to accumulate at the interface between the water and NAPL, and between water and air. The accumulation of the surface-active compound at the interface will

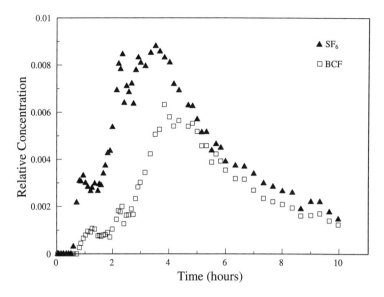

Figure 3.   Breakthrough curves measured for gas-phase transport of SF_6 (nonreactive) and BCF (partitioning) tracers in the vadose zone of a petroleum-contaminated site; from Bronson et al. (27).

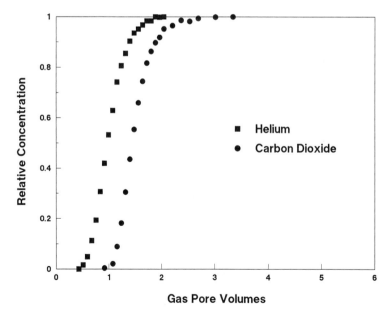

Figure 4.   Breakthrough curves measured for gas-phase transport of helium (nonreactive) and CO_2 (partitioning) tracers in a water-unsaturated column; from Brusseau et al. (28).

retard its transport with respect to that of a conservative tracer in a manner comparable to that of a bulk-phase partitioning tracer. Hence, the retardation of the surface-active compound will not only be a function of partitioning to the bulk phase but also to the interface. For this system, neglecting sorption by the solid phase, the retardation factor is defined as:

$$R = 1 + \frac{S_n K_n}{(1-S_n)} + \frac{A_i K_i}{\theta_w} \tag{3}$$

where A_i = area of the interface, $K_i = \Gamma/C_w$, the equilibrium partition coefficient between the interface and water or gas, Γ = adsorption density (mass of contaminant at interface/area of interface), and C_w is aqueous concentration. Ideally, the interface-partitioning tracer would have low partitioning to the bulk phase and negligible sorption by the soil phase. With knowledge of the interface partition coefficient (K_i), and an estimate of the water content (e.g., from porosity and S_n measurements), the area of the interface can be calculated. It should be noted that this value is an effective area, reflecting the influence of system properties (solute behavior, fluid distribution, etc).

The interface partitioning tracer should not alter the bulk phase, such that the "natural" form of the system can be measured. The tracer could alter the bulk phase, for example, through formation of emulsions or reductions in interfacial tensions. In addition, the partitioning tracers should not cause a change in the mass of the bulk phase, such as by solubilization or evaporation. The application and potential limitations of interface partitioning tracers requires much additional research.

Case Study

Interwell partitioning tracer tests were conducted at operable unit one (OU1) in Hill AFB, Utah before and after the application of a cyclodextrin complexing sugar flush (13,31,32). The surficial aquifer at the site was heavily contaminated with a complex LNAPL mixture comprised of organic compounds ranging from petroleum products to chlorinated solvents. The tests were conducted in a 3 m by 5 m cell bounded by corrugated iron walls that were driven into an underlying clay unit. A line of injection wells and a line of extraction wells, both fully screened, were installed at opposite ends of the cell. The water table was maintained at 5.3 m below ground surface, creating a 3 m saturated thickness. The cumulative injection and extraction flow rates were maintained at 4.54 L/min, equivalent to approximately one pore volume per day.

The breakthrough curves for the various conservative and partitioning tracers were analyzed using the method of moments. Due to the effects of heterogeneities and rate-limited mass transfer upon tracer transport, the extended elution tails of these breakthrough curves proved to be significant. Thus, it was necessary to extrapolate them to "zero concentration" by fitting an exponential decay function to the final portions of the breakthrough curves using an approach adapted from Pope et al. (33). This technique produced the most consistent and reliable NAPL saturation values.

The use of moment analyses to determine S_n values without first applying this "tail correction" consistently underestimated the degree of NAPL saturation. Chemical and microbiological interferences have prohibited the analysis of several tracers for the accurate estimation of NAPL saturations. Ethanol, pentanol, and hexanol experienced a significant degree of biodegradation, causing serious overestimation (ethanol) and underestimation (pentanol, hexanol) of NAPL saturations. The most reliable tracer pair was bromide as the conservative tracer and 2,2-dimethyl-3-pentanol (DMP) as the partitioning tracer. This pair was also included in the post-PTT tracer suite and again proved dependable. The partitioning tracer, 6-methyl-2-heptanol, included only in the post-PTT, also provided robust results. The performance of these tracers within the experimental system allowed a single S_n to be determined per extraction well for the pre-PTT and three separate S_n estimates per extraction well for the post-PTT. The multiple S_n values were averaged arithmetically for each well, and a cell-wide S_n value was calculated by flow-weighting the individual well values.

As discussed above, these tests allowed a much larger volume of aquifer to be characterized relative to more traditional means of NAPL quantification. Accordingly, it should be noted that the NAPL saturation estimates derived from the tracer tests are reflective of the *entire* swept volume of the contaminated zone. Thus, the S_n values determined with the PTT are effective values that represent the entire cell pore volume (in the case of cell-averaged values). Localized portions of the test cell exhibited NAPL saturations both larger and smaller than the cell-averaged values.

The averaged NAPL saturation values are presented in Table 1. The results of the preliminary partitioning tracer test indicate an increasing NAPL saturation from extraction well 2451 to 2453 (east to west). This trend is also reflected in the results from the post-partitioning tracer study, although the difference in NAPL saturation from well to well is reduced due to significant removal of contaminant mass by the remediation system. The value of S_n varied somewhat with the tracer pair used in the calculations for the post-PTT. Local S_n values, determined from total organic carbon (TOC) analysis of soil core material collected during preliminary coring, ranged from 11.9% to 17.6% saturation. Note that as expected, these values bracket the cell-averaged value of 12.1% obtained from the preliminary PTT. Furthermore, the range of local S_n values determined by TOC analysis compares well with the range of pre-PTT values for individual extraction wells, 10.1% to 14.8%. These results indicate that the partitioning tracer method provided robust estimates of NAPL saturation.

The estimates of NAPL saturations obtained from the preliminary and final tracer tests were used to evaluate the performance (contaminant mass removal) of the cyclodextrin flushing technology. The flushing of the test cell with the cyclodextrin solution reduced the effective NAPL saturation by 42.0% (S_n reduced from 12.1% to 7%). This volume (or mass) removal percentage is remarkably similar to the 41.4% mass-removal value obtained independently from analysis of core samples collected before and after cell remediation (31,32). This is further indication that the partitioning tracer test provided an accurate measure of NAPL saturation within the cell.

Table 1. LNAPL Saturation Values (S_n).

Pre-Technology Partitioning Tracer Results

Well	Tracer Pair	S_n
2451	Bromide/DMP	10.1%
2452	Bromide/DMP	12.9%
2453	Bromide/DMP	14.8%
Cell Mean[1]		12.1%

Post-Technology Partitioning Tracer Results

Well	Tracer Pair	S_n
2451	Bromide/DMP	5.7%
	Bromide/MH	6.1%
	DMP/MH	6.6%
	Mean	6.1%
2452	Bromide/DMP	7.9%
	Bromide/MH	7.3%
	DMP/MH	6.5%
	Mean	7.2%
2453	Bromide/DMP	7.7%
	Bromide/MH	7.7%
	DMP/MH	7.8%
	Mean	7.7%
Cell Mean[1]		7.0%

---

[1] Mean values are weighted by the swept volume associated with each well.
Note: DMP = 2,2-dimethyl-3-pentanol, MH = 6-methyl-2-heptanol

In conclusion, the application of a partitioning tracer study for the evaluation of remediation effectiveness provided useful information regarding a) the volume of NAPL present, b) the amount of NAPL removed and c) the aquifer's hydraulic properties. Solute biodegradation, site heterogeneity, and mass transfer constraints prevented the use of some tracers and forced more intensive data analysis techniques. However, the final results were consistent with more traditional characterization methods.

Conclusions

The partitioning tracer method provides a measure of fluid saturation (NAPL, water) at a scale that is much larger than that associated with point-measurement methods. This is an advantage for economical detection and measurement of immiscible-liquid saturation for large areas. The partitioning tracer concept can be applied to the measurement of interfacial areas, as well as bulk fluid saturations. As discussed by Brusseau et al. (28), using a suite of phase-selective partitioning tracers in combination with nonreactive tracers may be a viable and valuable approach for characterizing contaminated sites. The partitioning tracer method can be used to assist in the identification of NAPL source zones, to help evaluate the performance of remediation systems, and as a source of information for risk assessments. The method has recently been used successfully at several sites in the U.S.A. While the method has great promise, its performance can be influenced by many factors. Thus, the results obtained from a partitioning tracer test should be used with caution. Additional research, at both the laboratory and field scale, is needed to more fully evaluate the efficacy of the method.

Acknowledgements

This work was supported by grants provided by the U.S. Environmental Protection Agency and the Environmental Management Science Program of the U.S. Department of Energy.

Literature Cited

1. National Research Council. 1994. Alternatives for Ground Water Cleanup. National Academy Press, Washington, D.C.
2. Martin, A.J.P.; Synge, R.L.M. 1941, Biochem. J. 35, 1358-1368.
3. Cooke, C.E. 1971. U.S. Patent Number 3,590,923.
4. Deans, H.A. 1971. U.S. Patent Number 3,623,842.
5. Tang, J.S. 1995. SPE Formation Evaluation. 33.
6. Bouchard, D.C.; Enfield, C.G.; Piwoni, M.D. 1989. In: Reactions and Movement of Organic Chemicals in Soils, SSSA Spec. Publ. 22, 349-371.
7. Brusseau, M.L. 1990. In: Proc. Inter. Conf. Transport and Mass Exchange Processes in Sand and Gravel Aquifers, Atomic Energy Canada.

8. Brusseau, M.L. 1992. Water Resour. Res. 28, 33-45.
9. Jin, M.; Delshad, M.; Dwarakanath V.; McKinney D.C.; Pope G.A.; Sepehrnoori K.; Tilburg C.E. 1995. Water Resour. Res., 31, 1201-1211.
10. Wilson, R.D.; Mackay D.M. 1995. Environ. Sci. Technol. 29, 1255-1258.
11. Bedient, P.B.; Holder, A.W.; Enfield, C.G.; Wood, A.L. 1998. In: **Field Testing of Innovative Subsurface Remediation and Characterization Technologies**, Brusseau, M.L.; Sabatini, D.; Gierke, J., eds. American Chemical Society, Wash. D.C. (this volume).
12. Annable, M.; Rao, P.S.C.; Hatfield, K.; Graham, W.; Wood, L. 1998. Environ. Engin. (in press).
13. Cain, R.B.; Johnson, G.; Blanford, W.; Brusseau, M.L. 1996. (unpublished data).
14. Nelson, N.T.; Brusseau, M.L. 1996. Environ. Sci. Technol., 30, 2859-2863.
15. Chen, L.; Knox, R.C. 1997. Groundwater Monit. Remed., Summer, 161-168.
16. Gelhar, L.W.; Welty, C.; Rehfeldt, K.R. 1992. Water Resour. Res., 28, 1955-1974.
17. Miller, C.T.; Poirier-McNeill, M.M.; Mayer, A.S. 1990. Water Resour. Res. 26, 2783-2796.
18. Mayer, A.S.; Miller, C.T. 1992. J. Contam. Hydrol., 11, 189-213.
19. Powers, S.E.; Abriola, L.M.; Weber, W.J. 1992. Water Resour. Res. 28, 2691-2705.
20. Powers, S.E.; Abriola, L.M.; Weber, W.J. 1994. Water Resour. Res. 30, 321-332.
21. Brusseau, M.L.; Piatt, J.J.; Blanford, W.P.; Wang, J.; Hu, Q. 1998 (in review).
22. Srivastava, R.; Brusseau, M.L. 1998 (in review).
23. Tang, J.S.; Harker B. 1991. J. Can. Pet. Technol. 30, 76.
24. Tang, J.S.; Harker B. 1991. J. Can. Pet. Technol., 30, 34.
25. Whitley, G.A.; Pope, G.A.; McKinney, D.C.; Rouse, B.A.; Mariner, P.E. 1995. In: Monitoring and Verification of Bioremediation, Batelle Press, 211-221.
26. Simon, M.; Brusseau, M.L.; Golding, R.; Cagnetta, P.J. 1998. In: Proc. Inter. Conf. Remediation of Chlorinated and Recalcitrant Compounds, May 18-21, Monterey, CA, Battelle.
27. Bronson, K.; Nelson, N.T.; Brusseau, M.L. 1998. Unpublished data.
28. Brusseau, M.L.; Popovicova, J.; Silva, J. 1997. Environ. Sci. Technol., 31, 1645-1649.
29. Nelson, N.T.; Brusseau, M.L.; Silva, J. 1997. Presented at the National Meeting of the Amer. Geophysical Union, May 27-30, Baltimore MD.
30. Saripalli, K.P.; Kim, H.; Rao, P.S.C.; Annable, M.D. 1997. Environ. Sci. Technol., 31, 932-936.
31. Brusseau, M.L.; McCray, J.E.; Johnson, G.R.; Wang, X.; Wood, A.L.; Enfield, C. 1998. In: **Field Testing of Innovative Subsurface**

Remediation and Characterization Technologies, Brusseau, M.L.; Sabatini, D.; Gierke, J., eds. American Chemical Society, Wash. D.C. (this volume).

32. McCray, J.E.; Brusseau, M.L. 1998. Environ. Sci. Technol. 32, 1285-1293.

33. Pope, G. A. and K. Sepehrnoori, Final Report: NAPL partitioning interwell tracer test at OU1 test cell at Hill AFB, Utah, for work performed for ManTech Env. Res. Services Corp., P.O. #RC0251, GL #2000-602-46 00, Oct. 12, 1994.

Chapter 16

Investigation of Partitioning Tracers for Determining Coal Tar Saturation in Soils

Nancy J. Hayden and Harold C. Linnemeyer[1]

Department of Civil and Environmental Engineering, University of Vermont, Burlington, VT 05405

The use of partitioning tracers for determining NAPL saturation at a coal tar site was investigated. The study involved: 1) screening alcohol tracers for their appropriateness with coal tar; 2) conducting column tracer studies; and 3) conducting a partitioning interwell tracer test using existing site wells. The two alcohols having consistent partition coefficients were 4-methyl-2-pentanol and 2,4-dimethyl-3-pentanol. Column study results showed that coal tar saturation determined using the tracers compared favorably to known values and those determined using the soxhlet extraction method. Results from a flow rate study suggested that mass transfer limitations need to be considered. The partitioning interwell tracer test was conducted at two locations on the site. In one test, the partitioning tracers were not detected. At a second location, small separation in the tracer breakthrough curves was obtained resulting in low saturation estimates. These matched favorably with values from soil cores taken from the site.

The manufacturing of heating gas from coal which took place in the early part of the century, has left behind numerous sites contaminated with coal tar. Coal tar is a nonaqueous phase liquid (NAPL) composed of hundreds of constituents, such as polynuclear aromatic hydrocarbons, and volatile aromatic compounds. The actual chemical makeup can vary and is dependent on the type of gasification process, and raw materials used to manufacture the gas.

The compositional diversity of coal tar results in variations in its properties such as density and viscosity. The density can vary from slightly greater than to slightly less

[1] Current address: Intera Inc., 9111 Research Boulevard, Austin, TX 78758.

than that of water. Coal tar contamination, therefore, may be composed of both dense NAPLs (DNAPLs) and light NAPLs (LNAPLs) even at the same site. The viscosity of coal tar is typically much greater than water, with some coal tars exhibiting non-Newtonian behavior. High viscosity would result in slow subsurface movement of the coal tar NAPL. High viscosity coupled with the fact that coal tar has a very low overall solubility means that coal tar does not dissipate readily into the environment but can be a long-term source of soil and groundwater contamination.

Knowledge of the amount, and distribution of both the residual, and mobile phases of the coal tar is essential in order to determine the environmental and human health risk from the site and to determine an effective remediation strategy. Current methods for coal tar characterization are core sampling, cone penetrometer testing, and geophysical logging. The disadvantages of these methods are that they provide point measurements and sample only a very small portion of the contaminated area. The coal tar saturation can vary widely at a site depending on the nature and amount of the spill, the site geology, and the site hydrology. A soil measurement determined at a given point may not be very representative of the site as a whole unless a very large number of point measurements are taken.

Partitioning Tracer Methods

The partitioning interwell tracer test (PITT) is a method currently being investigated for determining NAPL saturation (1, 2, 3, 4, 5). The concept of using partitioning tracers for saturation determination was developed by the oil industry and has been used successfully in oil reservoirs (6, 7, 8).

The technique is based on the chromatographic separation of partitioning and nonpartitioning (conservative) tracers as they flow through the NAPL contaminated media. A partitioning tracer will distribute itself between the NAPL and water phase. The ratio of the concentration of tracer in the NAPL phase to the concentration of tracer remaining in the aqueous phase at equilibrium is called the partition coefficient (K).

A PITT is conducted by injecting both the partitioning and conservative tracer into an injection well, and determining their subsequent arrival time at a down gradient extraction well. As the partitioning tracer moves through the aquifer and partitions into the NAPL phase, a delay of the migration of the partitioning tracer relative to the nonpartitioning tracer occurs. If the partition coefficient is known, then the residual NAPL saturation can be determined from the tracer breakthrough curves, by the magnitude of this delay.

The delay of the partitioning tracer is referred to as retardation (R), and is equal to the time of arrival of the partitioning tracer (t_p) divided by the time of arrival of the nonpartitioning tracer (t_n). Two methods that are used to determine arrival time from the tracer breakthrough curves are; the method of moments, and the landmark comparison technique. The method of moments (used in our study) uses the centroid of the tracer breakthrough curves to determine t_p and t_n. These, in addition to the partition coefficient, are used to determine the NAPL saturation. The arrival time of the tracer is determined as the difference between the centroid of the tracer breakthrough curve and the centroid of the injected slug. If a tracer slug is injected at a constant flow rate and concentration, the centroid of the injected slug is equal to one-half the time of tracer slug injection. Additional details of this method are found in Jin et al. (2). Details of the landmark comparison technique are found in Tang and Harker (9,10).

For practical purposes, and ease of analysis some of the general conditions required for the use of partitioning tracer methods in environmental applications are that: 1) the flow through the aquifer be at or near steady state such that the partitioning and nonpartitioning tracers experience a similar flow rate; 2) the partition coefficient is relatively constant with respect to tracer concentration and NAPL saturation (over the range likely to be exhibited during the test); 3) equilibrium or near equilibrium conditions exist for the partitioning tracer between the aqueous phase and NAPL such that adequate characterization of the breakthrough curves is achieved; and 4) the retardation of the partitioning tracer is dominated by partitioning into the NAPL. Also important in environmental applications are that the tracers be relatively safe to humans and the environment, have an acceptable cost and be relatively easy to detect.

A PITT has the advantage of sampling a much larger area than the methods traditionally being used for NAPL characterization thus providing an average saturation of the aquifer volume sampled. A second advantage of this method is that the wells need not be drilled directly into the NAPL area. As long as the tracers pass through the contaminated area, average saturation estimates can be made.

The primary objective of this research was to investigate the use of the partitioning interwell tracer test (PITT) for characterization of coal tar saturation at a former coal gasification plant. The specific objectives were to: 1) identify a suitable partitioning tracer (or tracers) that could be used for coal tar in which the partition coefficient remains relatively constant with respect to alcohol concentration and coal tar saturation; 2) using laboratory studies, determine if the partitioning tracer test could adequately predict coal tar saturations in laboratory-packed columns of known coal tar saturation and field cores taken from the coal tar site; and 3) determine the applicability of using this method at a site using existing wells.

Methods

Partition Coefficient Determination. Seven alcohols were screened in an attempt to find an acceptable tracer for use with coal tar. These alcohols included pentanol, hexanol, heptanol, octanol, 2,2-dimethyl-3-pentanol, 2,4-dimethyl-3-pentanol, and 4-methyl-2-pentanol. A batch method was developed for determining the partition coefficient that provided a large surface area for mass transfer to occur but minimized mixing. Vigorous mixing was observed to cause coal tar emulsions. Glass beads (25 g, 3 mm diameter) were placed into the sample vial and coated with coal tar. Coal tar saturations of 25, 10, and 5 percent were obtained by adding 2.5 ml, 1.0 ml, or 0.5 ml of coal tar to the beads. The contents were then mixed on a reciprocating shaker for five minutes to distribute the coal tar as a film on the beads. The vials were filled with tracer solutions of various concentrations and mixed on a rotary mixer (~2 rev/min) for 24 hours. The samples were centrifuged to separate the water and coal tar phases. The tracer concentration was measured in the aqueous phase using a gas chromatograph with a flame ionization detector (GC/FID). Specific details of the analysis are presented in Linnemeyer (11). The partition coefficient shown in Equation 1 is the ratio of the concentration of tracer in the coal tar (C_{ct}) to the concentration in water (C_w) at equilibrium. Attempts to measure the tracer in the coal tar phase were abandoned due to the interference of coal tar peaks. The partition coefficient (K), therefore, was calculated from the water phase tracer concentrations and the difference between initial and final equilibrium concentrations shown in Equation 1.

$$K = \frac{C_{ct}}{C_w} = (\frac{C_{iw}}{C_w} - 1)\frac{V_w}{V_{ct}}$$

(1)

Where: C_{iw} is the initial aqueous phase tracer concentration;
V_w is the volume of water; and
V_{ct} is the volume of coal tar.

Column Tests. The partitioning tracer test method was performed on two types of soil columns: intact soil cores taken from a coal tar site; and laboratory-packed columns. Laboratory columns (glass with stainless steel endcaps) were packed with a dry sandy soil. The soil was removed and mixed with a known amount of coal tar and repacked. The column was water saturated and water in the up-flow direction was maintained for several days prior to testing.

The soil core samples were taken using 5 ft x 1.75 in (diameter) aluminum sample tubes that also served as soil columns in the laboratory. This provided an undisturbed sample for testing. Aluminum end caps fit the aluminum tubes by making a groove in it to hold a viton o-ring, thereby, providing a water tight seal.

The soil core sampling was conducted at three locations near the central portion of the site thought to be highly contaminated with coal tar. Soil cores were taken in the saturated zone from the water table to a confining silt layer. A hole was bored to the water table and once there, a sampling tube was installed and a core was taken. The tubes were sealed and stored at 4°C until testing. The columns were cut to the desired length, generally between 6 and 10 cm, in the laboratory when needed. Figure 1 shows a schematic of the field cores taken and and approximate depths of the portions used in the laboratory column studies.

The tracer solution in the columns consisted of approximately 10,000 mg/L bromide (conservative tracer), and 320 mg/L of 4-methyl-2-pentanol, and 2,4-dimethyl-3-pentanol (partitioning tracers). Relatively high concentrations were used to aid in the analysis using a bromide selective electrode. Flow was in the upward direction to minimize the potential density issues. The high concentration of bromide was not believed to pose a problem. Flow rates of 1 ml/min (Darcy velocity=1 m/d) and 0.2 ml/min (Darcy velocity=0.2 m/d) were used with the injection of a 30 ml slug of the tracer solution. Pore volumes varied slightly between columns, but were approximately 35 ml. Samples were taken every 15 minutes and analyzed using a bromide selective electrode and GC/FID.

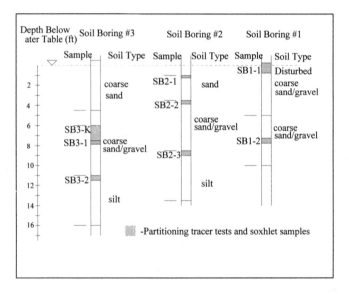

Figure 1. Field core approximate depths used for laboratory testing.

Field Tracer Tests. Tracer tests were initially conducted using bromide as a conservative tracer to evaluate potential wells for running a PITT. Bromide tests were first conducted on wells that showed promise based on the available site data such as hydraulic gradient and well screen intervals.

A partitioning interwell tracer test was conducted at two locations at the site that showed promise based on initial bromide tracer tests. A slug of tracer solution was prepared using river water and contained 78,100 mg/L bromide, 2,978 mg/L 4-methyl-2-pentanol, and 2,665 mg/L of 2,4-dimethyl-3-pentanol. The tracer was injected using a PVC pipe fit into the well casing. One end of the pipe was plugged using a rubber stopper with a line attached to the stopper. The pipe was placed into the well to the bottom and tracer solution was added to fill the pipe to the height of the water table. A weight was dropped to dislodge the plug, and the pipe was slowly pulled out of the well leaving behind the tracer solution distributed over the length of the well. This method of introduction resulted in a approximate tracer solution in the well of 39,000 mg/L bromide, 1500 mg/L 4-methyl-2-pentanol, and 1,300 mg/L of 2,4-dimethyl-3-pentanol. An ISCO Model 3700 portable sampler was used to obtain 300 ml samples at the extraction well (which is <10 % of the well bore) every two hours.

One test was performed between two nonpumping wells, with one well down gradient of the other. The wells were approximately 20 feet apart and the linear ground water velocity was calculated to be 2.2 m/d. Bromide concentration peaked between 64 to 70 hours after injection and with a resulting reduction in tracer concentration of about 3,100 times. The partitioning tracers could not be quantified in any of the samples.

It was apparent from these tests and some simple mathematical analysis that the groundwater was not flowing directly between these two wells but was veering away from the tracer extraction well. Only the outer portion of the plume was being sampled and hence the low tracer concentrations. This location was excluded from further study.

A PITT was performed at a second location consisting of a monitoring well (injection) that was approximately six feet from a groundwater pumping (tracer extraction) well. At this distance the injection well was within the capture zone. Although other monitoring wells were also located nearby, the injection well used was determined to be the best based on its screened interval, depth and radial distance from the extraction well. Due to the underlying silt layer about 12 to 15 below the water table and the pumping radius of influence, the salt concentration of the slug, at this location, was not considered to be a major concern.

Results

Partition Coefficient Determination. The implementation of a successful partitioning interwell tracer test (PITT) is dependent on the selection of a suitable partitioning tracer. The partitioning tracers chosen for the partitioning tracer test should have a partition coefficient (K) which is relatively constant with respect to coal tar saturation and tracer concentration. The partition coefficient should also be such that the retardation factor (R) falls within the range of 1.2 to 4. This ensures that the test can be conducted in a reasonable time frame while obtaining adequate separation of the breakthrough curves. The first task of this research was to screen different alcohols for use as a partitioning tracer with coal tar. The coal tar samples tested included a composite LNAPL sample taken from the site, and a DNAPL sample taken from a different site. For the final testing, an LNAPL taken from a monitoring next to the location of the soil cores was used and the additional DNAPL sample. No DNAPL was found in monitoring wells at this site. A summary of the results is shown in Table I.

Table I. Results of alcohols screened in this study. Saturations were 5, 10 and 25%.

Tracer	Coal Tar Sample	Comments
Pentanol	Composite	K small, K varied with concentration and saturation.
	DNAPL	
Heptanol	Composite	K too large, did not test over range.
	DNAPL	
Octanol	Composite	K too large, did not test over range.
Hexanol	Composite	K varied from 5-12 for concentrations ranging from 52-5200 mg/L and also varied with saturation.
	DNAPL	
2,2-dimethyl-3-pentanol	DNAPL	K varied slightly from 22 to 17 with saturation from 5-25%, concentrations used 3300 mg/L, 340 mg/L, expensive.
2,4-dimethyl-3-pentanol	DNAPL	K fairly consistent and within required range, concentrations used 12 mg/L and 340 mg/L. K varied with coal tar sample used.
	LNAPL site	
4-methyl-2-pentanol	DNAPL	K fairly consistent and within require range. Concentrations 12 and 100 mg/L. K varied with coal tar sample used.
	LNAPL site	

Tracers which had retardation factors outside the range of 1.2 to 4 were disregarded after the initial screening stages. The results of the hexanol tests showed that the partition coefficient varied with respect to both initial hexanol concentration and coal tar saturation. At low coal tar saturations (5%), the initial hexanol concentration had a much more pronounced effect on partition coefficient variability than at high coal tar saturations with K varying from 2.6 to 14 for initial hexanol concentrations from 5-5000 mg/L. At high coal tar saturation (25%), K varied from 9.2 to 10.0 for the DNAPL sample and initial hexanol concentrations from 52-5200 mg/L. Raimondi and Torcaso (12) and Knaepen et al. (13) also noted concentration effects on the partition coefficient in oil/brine systems.

The variation in the partition coefficient results were initially thought to be due to the batch testing method even though the vigorous mixing was eliminated. Luthy et al. (14) suggested that inadvertent sampling of coal tar water emulsions were the cause of high alcohol concentrations being measured in the water phase of their alcohol/coal tar partitioning experiments. Gschwend et al. (15) noted that non-settling particles and organic macromolecules could affect the octanol water partition coefficient value when using the batch method.

In order to determine if the concentration effect was due to the batch testing method used, partition coefficients were determined using clean sand-packed columns of known coal tar saturation. One column had a saturation of 6.3% and the other a saturation of 9.6%. Two tracer concentrations were tested in each column. Partition coefficients calculated from the soil column with 9.6% saturation were 5.2 and 9.1 for hexanol concentrations of 680 and 5000 mg/L, respectively. Partition coefficients calculated from the soil column with 6.3% saturation were 5.4 and 12.7 for hexanol concentrations of 630 and 5400 mg/L, respectively. Clearly, a strong concentration effect was evident. Hexanol was deemed to be an unsuitable coal tar partitioning tracer.

The two tracers deemed suitable were 4-methyl-2-pentanol and 2,4-dimethyl-3-pentanol. The average K values for three coal tar samples are shown in Table II. These values represent the average of three replicates for each saturation (5, 10, and 25%) and two initial tracer concentrations (27 and 340 mg/L for sample 1, and 12 and 100 mg/L for sample 2). Sample 3 is the average of 3 replicates for one concentration (27 mg/L) and one saturation (10%). An accurate K could not be determined for sample 3 for 4-methyl-2-pentanol, because the coal tar peaks interfered with the 4-methyl-2-pentanol peak. These results show some variation of K with respect to the coal tar sample and thus the importance of using a representative sample from the site for partition coefficient determination.

Table II. Average K values for tracers at varying saturations and concentrations.

Sample	Coal Tar	4-methyl-2-pentanol K (standard deviation)	2,4-dimethyl-3-pentanol K (standard deviation)
1	LNAPL site	4.1 (0.36)	21.4 (2.96)
2	DNAPL	3.0 (0.95)	18.1 (1.23)
3	DNAPL	Peak interference	27.7 (4.65)

Laboratory Column Studies. To evaluate the performance of the partitioning tracer test for predicating coal tar saturations, a series of partitioning tracer tests were conducted in both; laboratory-packed columns of known coal tar saturations, and soil cores taken from the field. The coal tar saturation determined from the partitioning tracer test was compared to the soxhlet extraction measured value. Soxhlet extraction involves extracting the sample for 18 hours using a solvent, dichloromethane in our case (EPA method 3540).

Columns which had been prepared in the laboratory and saturated with 9.6% and 15% coal tar were used to conduct partitioning tracer tests. Tests were run in the 9.6% column using flow rates of 1 ml/min and 0.2 ml/min to determine the effect of flow rate on the partitioning tracer results and saturation estimates. A flow rate of 1 ml/min was chosen because it corresponded to a linear velocity of about 2.8 m/d. Linear velocities estimated at the site varied between 2.2 m/d and 3.6 m/d from bromide tracer tests performed at the site.

The calculated saturation values at the higher flow rate of 6.4% for 4-methyl-2-pentanol and 5.8% for the 2,4-dimethyl-3-pentanol were significantly lower than the true value of 9.6 percent. At the lower flow rate, the values estimated from the partitioning tracer method increased significantly compared to those determined at the higher flow rate (14.5% for 4-methyl-2-pentanol and 8.5% for the 2,4-dimethyl-3-pentanol) suggesting that mass transfer limitations at the higher flow rate may affect the saturation estimate. The lower flow rate, therefore, was used in all subsequent column tests. The predicted saturation values for the 15% column were 14% for 4-methyl-2-pentanol and 11.2% for the 2,4-dimethyl-3-pentanol determined at the 0.2 ml/min flowrate. Figure 2 shows the breakthrough curves for this 15% column. The tail of the curves beyond the final sample was modeled using an exponential decay (3).

Laboratory partitioning tracer tests were run using subsections taken from the site soil cores shown earlier in Figure 1. These tests were conducted to further assess the effectiveness of the partitioning tracer method. The coal tar saturations determined using the partitioning tracer test were then compared to saturations determined from soxhlet extractions. The field cores were cut from various depths to collect information on the vertical distribution of the coal tar contamination at the site.

The breakthrough curves for the partitioning tracer test run on site cores showed little separation of the partitioning and nonpartitioning tracers peaks, with the separation evident primarily in the tail of the curves. Five columns taken from the field cores were tested using the partitioning tracer test and of these, three were also soxhlet extracted. Two more subcores were soxhlet extracted only. The estimated coal tar saturations in these seven cores were all less than 6%. A comparison of saturation estimates determined from the partitioning tracer method and soxhlet extractions for the laboratory-packed columns and field cores is shown in Figure 3. The one to one line is also shown. The coal tar saturations predicted by soxhlet extraction ranged from 0.8% to 2.25% for the field cores. The 4-methyl-2-pentanol predicted saturations from 1.0% to 5.4% and the saturations estimated by the 2,4-dimethyl-3-pentanol were in the range of 0.8% to 2.0% for the cores tested.

Generally, the estimates using the 2,4-dimethyl-3-pentanol tracer matched the estimates of the soxhlet extraction method. This indicates that the 2,4-dimethyl-3-pentanol can estimate the coal tar saturation with the accuracy of the currently used soxhlet extraction method. Unfortunately, the solvent extraction method for coal tar using dichloromethane is at best 60-70% efficient in extracting the coal tar from the soil (16). The data also reveal that in the tests performed in the columns of known saturation

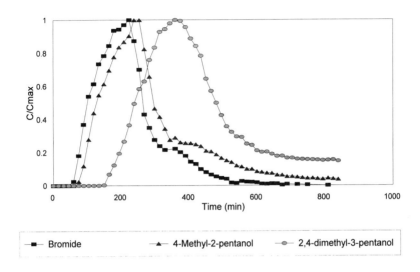

Figure 2. Breakthrough curves from a 15% saturated column.

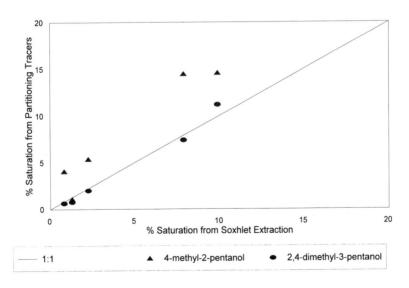

Figure 3. Comparison of tracer and soxhlet results.

the soxhlet extraction and the 2,4-dimethyl-3-pentanol predicted saturations that were less than the known saturation.

The saturation estimates determined using 4-methyl-2-pentanol were consistently higher than the soxhlet and 2,4-dimethyl-3-pentanol estimates. 2,4-dimethyl-3-pentanol is a larger molecule than 4-methyl-2-pentanol, and will have a slower diffusion rate through both the water and NAPL phases. From a practical standpoint, it is possible that the lower saturation estimates of 2,4-dimethyl-3-pentanol result from mass transfer limitations of the tracer, suggesting the need for an appropriate flow rate during the test.

Field Tracer Tests. The main objective of this project was to demonstrate the use of the partitioning interwell tracer test (PITT) for use as an in-situ method to determine residual coal tar saturations. The final phase of this research, therefore, was to conduct a successful PITT at the coal tar site.

Figure 4 shows the breakthrough curves for this field test. Bromide was detected about four hours after the tracer slug was injected, and the peak concentration, of about 130 mg/L, occurred seven hours after the start of the test. The partitioning tracers peaked at the same time as the bromide, and there was no difference between the curves until about twelve hours after the start of the test. This separation pattern was similar to the breakthrough curves obtained from the partitioning tracer tests conducted in the laboratory using field cores. In the laboratory trials, there was also very little retardation of the peaks, with the lag typically exhibited in the falling limbs of the breakthrough curves.

The method of moments was used to determine the coal tar saturation for the site test using the partitioning tracers. From the 4-methyl-2-pentanol curve the estimated coal tar saturation was 1.9%. Using the 2,4-dimethyl-3-pentanol curve a predicted saturation of 2.9% was obtained. The partition coefficients determined for the coal tar sample taken from the site were used for these calculations.

The results from the PITT agree well with the laboratory partitioning tracer tests, and the soxhlet extractions performed on the field cores suggesting that the partitioning tracer test method can give reasonable predictions of residual coal tar saturations. The small retardation of the partitioning tracers was expected based on the laboratory results.

The high groundwater velocity encountered in the test area (approximately 3.6 m/d) could also have contributed, in part, to the low coal tar saturation estimates. The laboratory-packed soil column tests showed that the higher velocities resulted in lower saturation estimates for both the 4-methyl-2-pentanol and 2,4-dimethyl-3-pentanol. This may have been a problem in the field test as well.

Summary and Conclusions

The primary objective of this study was to investigate the use of the partitioning interwell tracer test (PITT) for determining coal tar saturations at the site of a former coal gasification plant. The first phase of this research involved the selection of partitioning tracers that would be suitable for use with coal tar having partition coefficients that were constant with respect to coal tar saturation and tracer

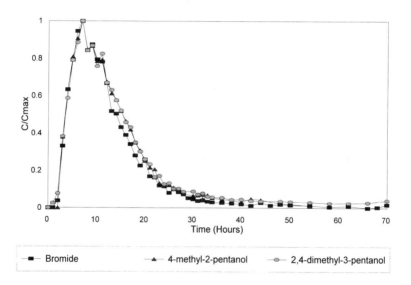

Figure 4. Breakthrough curves from field test.

concentration. Two alcohols 4-methyl-2-pentanol, and 2,4-dimethyl-3-pentanol were found to meet the tracer selection criteria and were used as partitioning tracers for the soil columns and field tests. The partition coefficients for these two tracers, however, were found to vary with respect to the coal tar sample used for the analysis. The partition coefficient of the 4-methyl-2-pentanol was found to be in the range of 3.6 to 4.1, and the K for the 2,4-dimethyl-3-pentanol varied from 18.7 to 27.4 depending on the coal tar sample. A coal tar sample representative of the site should be used for the K determinations to obtain an accurate partition coefficient.

Laboratory studies were conducted to determine if the partitioning tracer test could adequately predict coal tar saturations in laboratory-packed columns of known coal tar saturations and field cores taken from the site. The coal tar saturations predicted by soxhlet extraction ranged from 0.8% to 2.25% for the field cores. The 4-methyl-2-pentanol predicted saturations from 1.0% to 5.4% and the saturations estimated by the 2,4-dimethyl-3-pentanol were in the range of 0.8% to 2.0%. The results showed that the partitioning tracer test could provide reasonable estimates of coal tar saturation. The saturation predicted from the partitioning tracer tests performed in the 9.6% coal tar saturated columns increased when the flow rate was decreased indicating that equilibrium had not been achieved at the higher flow rate. The groundwater flow rate may affect saturation estimates when conducting a PITT at coal tar sites.

The final phase of the study was to conduct a PITT at a coal tar site to determine the applicability of the test for use with existing site wells. The partitioning interwell tracer test performed at the coal tar site provided coal tar saturation estimates which were consistent with values obtained from the field cores. The results showed that the PITT can be successfully applied to existing sites if the issues of flow rate and tracer dilution are addressed. The results from this research suggest that the partitioning interwell tracer test could be a useful method for characterization of coal tar saturations.

Acknowledgments

This research was supported by the Vermont Department of Environmental Conservation. Special thanks to Mr. Richard Spiese, Mr. Jim Bowes of The Johnson Company, and Mr. Kenneth Zegel.

Literature Cited

(1) Annable M.D.; Rao, P.S.C.; Hatfield, K.; Graham, W.D.; Wood, A.L. *Proceedings from the 2nd Tracer Workshop, Austin TX*, **1995**.
(2) Jin, M.; Delshad, M.; Dwarakanath, V.; McKinney, D.C; Pope, G.A.; Sepehrnoori, K.; Tilburg, C.E. *Water Resources Research*. **1995**, 31, 1201.
(3) Pope, G.A.; Jin, M.; Dwarakanath, V.; Rouse, B.; Sepehrnoori, K. *Proceedings from the 2nd Tracer Workshop, Austin TX*, **1995**.
(4) Wilson, R.D.; Mackay, D.M. *Environ. Sci.Technol.* **1995**, 29, 1255.
(5) Nelson, N.T.; Brusseau, M.L. *Environ. Sci.Technol.* **1996**, 30, 2859.
(6) Lichtenberger, G.J. *Production Operations Symposium.* **1991**, SPE 21652.
(7) Tang J.S. *Journal of Canadian Petroleum Technology.* **1992**, 31, 61.

(8) Tang J.S. *Society of Petroleum Engineers Formation Evaluation*. **1995**, 3, 33.

(9) Tang J.S.; Harker, B. *Journal of Canadian Petroleum Technology*. **1991**, 30, 76.

(10) Tang J.S.; Harker, B. *Journal of Canadian Petroleum Technology*. **1991**, 30, 34.

(11) Linnemeyer, H.C. MS Thesis, University of Vermont, 1997.

(12) Raimondi P.; Torcaso M. *SPE Journal*. **1965**, 5, 51.

(13) Knaepen, W.A.I.; Tjssen, R.; van den Bergen, E.A. *SPE Reservoir Engineering*. **1990**, 5, 239.

(14) Luthy, R.G.; Dzombak, D.A.; Peters, C.A.; Roy, S.B.; Ramaswami, A.N.; Nakles, D.V.; Nott, B.R. *Environ. Sci. Technol*. **1994**, 28, 266.

(15) Gschwend, P.M.; Wu, S.C. *Environ. Sci. Technol*. **1985**, 19, 90.

(16) Guillen, M.D.; Blanco, J.; Canga, J.S.; Blanco, C.G. ACS Energy and Fuels. **1991**, 5, 188.

Chapter 17

A Biotracer Test for Characterizing the In-Situ Biodegradation Potential Associated with Subsurface Systems

Mark L. Brusseau[1,2], Joseph J. Piatt[1], Jiann-Ming Wang[1], and Max Q. Hu[1]

[1]Departments of Soil, Water, and Environmental Science and [2]Hydrology and Water Resources, University of Arizona, Tucson, AZ 85721

Evaluating the feasibility of using intrinsic or accelerated in-situ bioremediation for a specific site requires a determination of the in-situ biodegradation potential of the target contaminants in the contaminated zone, which is a very difficult task. The purpose of this paper is to introduce a field-scale, controlled-release approach based on the use of a biodegradation tracer test. This method entails conducting a tracer experiment with one or more compounds whose biodegradation characteristics are well known. The biotracer test can be used to: (1) evaluate the general biodegradation potential associated with the zone of interest, (2) evaluate the response of the system to perturbations such as oxygen addition, and (3) evaluate the biodegradation potential for a specific contaminant. The utility of the biotracer test for the first application is illustrated with a test conducted at a field site contaminated by jet fuel. The results of the experiments indicate that it is possible to characterize the degradation potential of a selected site using biotracers. This method may, therefore, be a useful addition to our arsenal of methods for evaluating the feasibility and performance of in situ bioremediation.

The biodegradation potential of organic compounds in the subsurface has long been of interest to environmental and soil scientists. For example, the efficacy of a pesticide depends, in part, on its resistance to biodegradation in the target zone. The possibility of a pesticide or industrial compound contaminating groundwater also depends, in part, on the biodegradability of the compound. The advent of in-situ bioremediation as a preferred method for cleaning up contaminated sites has greatly increased interest in the biodegradation of organic compounds in the subsurface. This interest has recently been compounded by the consideration of natural attenuation via intrinsic bioremediation as a clean-up alternative.

Evaluating the feasibility of using intrinsic or accelerated in-situ bioremediation for a specific site requires a determination of the in-situ biodegradation potential of the target contaminants in the contaminated zone. Additionally, the design and performance-evaluation of in-situ bioremediation programs requires quantitative information concerning the magnitude and rate of expected and actual biodegradation.

As discussed by several authors (e.g., 1-3), it is very difficult to accurately determine the potential for biodegradation and bioremediation of a specific contaminant at a specific field site. It is even more difficult to quantify the magnitude and rate of biodegradation of the contaminant.

An approach often used to characterize biodegradation potential is to conduct laboratory experiments using soil samples collected from the field. While this approach is very useful for evaluating the potential for biodegradation in principle, the complexity of field sites often precludes transferring laboratory results directly to the field. For example, biodegradation is very sensitive to environmental conditions (identity and concentration of dominant electron acceptor, nutrients, pH, temperature, co-contaminants) and it is very difficult to fully reproduce field conditions in the laboratory. Thus, results obtained from laboratory experiments may not accurately reflect in-situ behavior. In addition, the representativeness of the samples collected from the field is uncertain, given the heterogeneity intrinsic to field sites.

The problems inherent to laboratory based methods can be partially resolved by using field-based methods. These methods typically involve monitoring temporal and spatial changes in concentrations of parameters associated with biodegradation, such as CO_2, O_2, intermediary metabolites, and the substrate (contaminant) itself. However, the complexity of field sites makes it very difficult to conclusively correlate changes in these parameters to biodegradation. For example, mass transfer (sorption, dissolution, volatilization) and abiotic transformation (hydrolysis, abiotic reduction) processes can influence contaminant transport and fate, thereby confounding an analysis of biodegradation. Furthermore, the initial mass of contaminant released into the subsurface is not known at most sites. Thus, it is not possible to conduct a mass balance, which makes it difficult to quantify the magnitude and rate of biodegradation.

Despite the problems discussed above, information regarding the biodegradation potential of a site is critical to the planning and implementation of biodegradation-based remediation systems and, consequently, efforts to obtain such information continue. The application of isotope analysis (2) and biomolecular methods (4) for characterizing field-scale biodegradation potential is under investigation. These methods, while promising, are based on the use of samples collected from the field and, as such, suffer some of the same constraints discussed above. A field-based method that has promise involves the use of "in-situ microcosms", a method used for some time in soil science and more recently in environmental science (5,6). A major characteristic of this method is that it provides data for a very localized zone (in the range of tens of centimeters). This is a positive attribute for purposes of studying the spatial variability of bioactivity. However, a representative sampling of an entire system may require conducting a time- and cost-prohibitive number of individual experiments.

Controlled-release field experiments conducted in the zone of interest have the greatest potential for providing definitive measures of the potential, magnitudes, and rates of biodegradation. The ultimate controlled-release experiment involves injecting the actual contaminant of interest into the target zone, which is the approach used in several previous research projects (e.g., 7-9). This approach, while ideal, will rarely be possible at most contaminated sites due to both technical (differentiating exogenous and resident contaminants) and regulatory (resistance to use of hazardous compounds) constraints. Alternatives discussed to date involve the use of electron acceptors, whose mass loss is considered an indicator of biological activity (2). As discussed previously, however, it is often difficult to quantify biodegradation based on the behavior of electron acceptors.

The purpose of this paper is to introduce a field-scale, controlled-release approach based on the use of a biodegradation tracer test. The biotracer test can be used to: (1) evaluate the general biodegradation potential associated with the zone of interest, (2) evaluate the response of the system to perturbations such as oxygen addition, and (3) evaluate the biodegradation potential for a specific contaminant. The method entails conducting a tracer experiment with one or more compounds whose biodegradation characteristics are well known. The transport and recovery of the biotracers are compared to those of a nonreactive tracer to evaluate and quantify the biodegradation potential for the target zone. For illustration, the method is applied to a field site contaminated by jet fuel.

THEORY
Basis of the Method

The three major problems limiting the in-situ characterization of biodegradation potential are: (1) difficulty in distinguishing between abiotic and biotic causes of mass loss, (2) initial mass and distribution of contaminant are unknown, and (3) uncertainty associated with heterogeneity. The design of the biotracer-test method addresses these problems. First, the injection and recovery of a known mass of substrate provides an opportunity for conducting mass balances, which is critical for quantitative analyses of biodegradation. Second, the impact of hydraulic and abiotic-reaction processes on transport can be evaluated by use of one or more conservative (non-biodegradable) tracers, which allows the influence of biodegradation to be separated from that of other processes. Third, the influence of heterogeneity on transport and biodegradation is inherently incorporated in the results of the tracer experiment.

The biotracer test is implemented by conducting a tracer experiment using both biodegradable and non-biodegradable compounds. Aqueous concentrations of the tracers are monitored at one or more locations during the test to determine breakthrough curves. A moment analysis of the data is conducted to obtain mass recoveries and travel times. Lower mass recoveries of the biotracers compared to those of the conservative tracers is generally indicative of biodegradation, assuming that other potential sources of mass loss are accounted for with the conservative tracer. This comparative mass recovery analysis also provides a direct measure of the magnitude of biodegradation. Information on the rates of biodegradation can be obtained by analysis of the mass recovery/travel time relationship, and by the use of mathematical models.

The breakthrough curve measured at a specific monitoring location reflects the impact of processes occurring along the entire length of each and every flow path connecting the injection and monitoring points. Thus, they represent an integration across the zone influenced by the tracer test. As a result, the biodegradation data obtained from a biotracer test are reflective of microbial processes occurring everywhere within the test domain. Furthermore, the quantitative information, such as rate coefficients, obtained from a biotracer test are, therefore, composite values representative of the entire domain. The ability to obtain a representative profile of biodegradation activity across a large spatial scale is a distinct advantage of the biotracer method compared to core-based methods. Sampling at multiple locations can be done to examine the (larger-scale) spatial variability of biodegradation.

Application of the Method

The biotracer test can be used to: (1) evaluate the general biodegradation potential associated with the zone of interest, (2) evaluate the response of the system to perturbations such as oxygen addition, and (3) evaluate the biodegradation potential for a specific contaminant. As discussed above, the biotracer test can be used to characterize the potential, magnitude, and rate of biodegradation. The biotracer test can also provide information about the nature of biodegradation. For example, the profiles of the breakthrough curves can be examined to evaluate if biodegradation is linear (e.g., a first-order process with minimal bacterial growth) or nonlinear (e.g., characterized by a lag phase and measurable growth). Analysis of the mass recovery/travel time relationship for data collected at multiple locations can provide an additional means to evaluate the linearity of biodegradation.

The biotracer test is ideal for evaluating the impact of perturbations on the system, the second application mentioned above. For example, the impact on biodegradation of adding an electron acceptor or nutrient to the system can be evaluated. This can be incorporated into the pilot-scale tests conducted during the initial stages of planning an in-situ bioremediation system.

The first two uses of the biotracer test are relatively straightforward. The third use of the biotracer test is, however, more uncertain. To evaluate the potential for, and the magnitude of, biodegradation of a specific contaminant, the degradation pathways and the degrading populations for the biotracer should be as similar as possible to those for the contaminant. As previously discussed, it is rarely possible to use the contaminant of interest in a tracer experiment. The next best choice is a low toxicity compound that has biodegradation characteristics similar to those of the contaminant. This can be accomplished by selecting an analogue or intermediary metabolite specific to the target contaminant.

For example, the aerobic degradation pathways for benzoate are similar to those of other aromatic compounds. Thus, benzoate could be used, in some cases, as a representative analogue for aerobic biodegradation of contaminants such as benzene, toluene, and xylene. In addition, benzoate is considered a model for anaerobic biodegradation of aromatic compounds (10-12). Thus, benzoate would be a good choice to evaluate the overall biodegradation potential for a petroleum-contaminated site, given that the aromatic compounds are generally of greatest environmental concern.

Implementation of the Method

The choice of biotracer should reflect the specific characteristics of the target site and the specific objectives of the test. For example, a biotracer that is readily degraded by many different types of bacteria would be used to evaluate the general biodegradation potential for a site. Conversely, to determine the biodegradation potential for a specific contaminant, a biotracer specific to the target contaminant would be used.

The choice of biotracer should, of course, reflect the redox status of the system. This can be illustrated using a petroleum-hydrocarbon contaminated site as a hypothetical target system. The concentration of oxygen in the water-saturated subsurface domains of many petroleum contaminated sites is relatively low due to the large supply of substrate. Thus, if one wished to determine the inherent biodegradation potential of the site, such as for evaluating the feasibility of intrinsic bioremediation, a biotracer would be selected that is biodegraded under low-oxygen conditions. Conversely, a biotracer that is biodegraded under aerobic conditions would be selected to represent conditions such as those associated with an accelerated in-situ bioremediation system involving oxygen addition. To characterize the general biodegradation potential, it would be desirable to use a biotracer that can be degraded under both aerobic and anaerobic conditions.

The toxicological characteristics of the biotracer should be such that its injection into the subsurface would be acceptable to regulatory agencies and community groups. The biotracer should also be relatively inexpensive. Benzoate is an example of a good candidate in both respects. It is an approved food additive, primarily as a food preservative, and, as such, is available in large quantities at relatively low cost.

The biotracer test is conducted similarly to a nonreactive tracer test, which are widely used to measure directions and rates of groundwater flow. Groundwater tracer tests can be implemented in a number of ways, based primarily on the number and configuration of the wells. The simplest configuration involves a single well, wherein the tracer solution is first injected into and then extracted from the aquifer (push-pull test). Water samples for analysis of tracer concentrations are collected at the well. The next simplest design involves a single extraction well and a non-pumping injection well, where the passive well is used to introduce the tracer solution into the flow field created by the extraction well, which is also used for sampling. Another variation involves a single pumping injection well and one or more monitoring wells surrounding the injection well. The fourth basic design involves a combination of pumping injection and extraction wells. The utility and effectiveness of different well configurations for nonreactive tracer tests was evaluated by Gelhar et al. (13), who concluded that the dual injection-extraction well configuration provided the most reliable data.

Limitations

As discussed above, the biotracer test provides a measure of biodegradation potential across a relatively large scale. This is advantageous for large-scale applications, but it limits the information that can be obtained for a very localized zone. Analysis of biotracer test results can be complicated by the presence of competing substrates. Thus, detailed characterization of the system is required to assess the potential for competitive effects. A limitation associated with the third use of the biotracer test is that it may be difficult in some cases to find a biotracer whose biodegradation characteristics are very similar to those of the target contaminant, and that also meets the other criteria (low toxicity, etc). Fortunately, representative compounds appear to be available for some of the most widespread contaminant classes, such as petroleum-based compounds.

With the appropriate selection of the biotracer, there is a high probability of success for evaluating the potential and magnitude of biodegradation. However, there may be greater uncertainty associated with using a biotracer to accurately measure the rate of biodegradation, depending on which factor or process controls the overall rate of biodegradation. The global biodegradation rate is influenced by several factors and processes, including acclimation time (genetic adaptation, enzyme induction), population growth, rate of catabolism, bioavailability (accessibility and uptake of substrate), and the supply of electron acceptors and nutrients.

Of all the factors influencing the rate of biodegradation, bioavailability is perhaps the one whose impact is most difficult to simulate with a biotracer test. The influence of flow-related bioavailability processes (e.g., mass transfer between low and high permeability domains) on the rate of biodegradation may be relatively well simulated with the use of a biotracer that resides only in the aqueous phase (e.g., nonsorbing). This may not be true, however, if the degree of mass-transfer constraint is extreme. In such cases, the measured biodegradation potential may be representative primarily of the more permeable zones wherein the majority of flow, and biotracer transport, occurred.

The biodegradation rates obtained with a biotracer test may not be representative when chemical-related bioavailability processes (sorption/desorption, NAPL dissolution) control the rate of biodegradation. This potential constraint can be at least partially addressed, however, by using a biotracer that will partition between water and the sorbed or NAPL phases. The biodegradation rates obtained for a partitioning tracer should be relatively accurate if the mass transfer associated with the partitioning process is not significantly rate limited. If partitioning is significantly rate limited, the rates may not be accurate.

CASE STUDY

An experiment conducted at a jet-fuel contaminated site located at Hill Air Force Base, Utah was used to illustrate the application of the biotracer-test method, focusing on its use in evaluating the general biodegradation potential of the site. Benzoate was selected as a biotracer representative of the biodegradation of alkylbenzenes. Salicylate, an intermediary of naphthalene biodegradation, was used to represent the biodegradation of lower molecular weight polycyclic aromatic hydrocarbons. Ethanol was used to represent the biodegradation of lower molecular weight alkanes. These compounds are non-volatile and should experience minimal hydrolysis under the system conditions. In addition, preliminary experiments have shown that sorption of these tracers by the site aquifer material is negligible. Thus, mass loss as a result of abiotic processes is unlikely. Pentafluorobenzoate was used as the nonreactive tracer.

A series of laboratory batch and column experiments were conducted using groundwater and aquifer material collected from the site to examine the inherent biodegradability of the biotracers. A field experiment was conducted to examine the biodegradation of the biotracers during their transport through a contaminated portion of the aquifer.

The field site is EPA Operable Unit One at Hill AFB in Layton, Utah. The tracer test was conducted in a portion of an unconfined aquifer that consists of fine to coarse sand, interbedded with gravel and silt layers. The natural saturated thickness of this unit ranges from zero to 3 meters. A clay unit, the top of which is about 10 meters below ground surface, underlies the aquifer. Several waste disposal areas are located in the vicinity of Operable Unit One. Of particular concern are two chemical disposal pits that are no longer in operation. Large quantities of liquid wastes (primarily petroleum hydrocarbons and spent solvents) were disposed of and periodically burned in the pits. The tracer test was conducted adjacent to one of the chemical disposal pits.

A complex mixture of water-immiscible organic compounds exists at varying saturations within the test zone. The immiscible-liquid mixture, while containing chlorinated compounds (e.g., solvents, PCBs), is considered to be less dense than water as a whole. Due to seasonal and diurnal fluctuations of the water table, the immiscible liquid resides in a smear zone throughout the saturated interval. Analysis of cores and water samples collected from the site shows that a multitude of petroleum compounds are present at mg/L concentrations. Dissolved oxygen concentrations of approximately 1-3 mg/L were measured for groundwater in the test zone.

The experiment was conducted in an enclosed 3x5 m^2 cell emplaced in the subsurface. The cell was created by driving sealable-joint sheet-pile walls into the clay layer. The cell was created with minimal disruption of the subsurface located within the cell. A line of four fully screened injection wells and a line of three fully screened extraction wells, both normal to the direction of flow and located approximately 4 meters apart, were used to create a steady state flow field.

After steady state flow was established, a 0.2 pore-volume pulse of tracer solution was injected into the cell, followed by approximately ten pore volumes of tracer-free water. A total flow rate of approximately one pore volume per day was maintained throughout the experiment ($Q \approx 4$ l/min). Flow was monitored and adjusted to maintain the water table at 5.4 ± 0.15 meters below ground surface, giving a saturated thickness of 3 meters. The influent concentrations for the four compounds were 262 mg/l (pentafluorobenzoate), 432 (benzoate), 355 mg/l (salicylate), and 1107 mg/l (ethanol).

A three-dimensional, multi-level sampling array (12 locations, 5 depth intervals) connected to a vacuum system was used to collect depth-specific water samples. Depth-integrated samples were collected from the three extraction wells. Samples were collected with minimal headspace in polyethylene vials containing an aliquot of $HgCl_2$, which served as a preservative to prevent biodegradation. The samples were stored at 4 ^0C until analysis. Concentrations of benzoate, salicylate, and pentafluorobenzoate were determined using HPLC (Waters) with a UV-VIS spectrophotometer. Ethanol concentrations were determined by GC-FID.

RESULTS AND DISCUSSION

Plate counts, with benzoate as the sole carbon source, were performed using groundwater and aquifer material collected from the site. The results showed that bacteria capable of degrading benzoate exist at the site at levels ranging from 10^3 to 10^4 colony forming units per gram of dry soil. $^{14}CO_2$ was evolved during the benzoate mineralization experiments, indicating that benzoate can be mineralized by the resident bacterial populations.

Representative breakthrough curves for transport of pentafluorobenzoate and benzoate in a column packed with Hill aquifer material are shown in Figure 1. The biotracer clearly exhibits mass loss compared to pentafluorobenzoate. The mass loss can be attributed to biodegradation, based on the results obtained from complementary

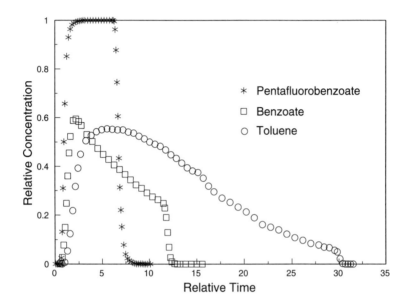

Figure 1.   Transport of tracers through a column packed with Hill aquifer material: breakthrough curves for pentafluorobenzoate (input pulse of 5.8 pore volumes), benzoate (10.9), and toluene (28.4).

column experiments conducted with non-acclimated aquifer material, as well as those from the batch experiments. The shapes of the breakthrough curves indicate that biodegradation is nonlinear for this system. Given these results, in combination with those obtained from the batch experiments, it appears that the selected compounds may successfully serve as biotracers for characterizing the general biodegradation potential at the Hill field site.

Breakthrough curves for pentafluorobenzoate, benzoate, salicylate, and ethanol were measured at several monitoring locations during the field experiment (see Figure 2 for an example). Mass recoveries for pentafluorobenzoate were approximately 100%, indicating mass conservation during the tracer experiment. Conversely, mass recoveries for benzoate, salicylate, and ethanol were less than 100%, as evidenced by comparing their breakthrough curves to those of pentafluorobenzoate.

The mass loss observed for the biotracers is attributed to biodegradation, given the characteristics of the biotracers, the absence of other loss mechanisms, and the results of the laboratory experiments. The degree of mass loss for the field experiment was in the range of 10 to 60%. The degree of mass loss observed for the data shown in Figure 2 is consistent with that specific sampling location.

The field test was designed to evaluate the use of the biotracers to measure the general biodegradation potential of the site. However, miscible-displacement experiments were conducted in the laboratory to compare the magnitude and rate of biodegradation of the biotracers to that of representative target contaminants. The transport of toluene, a representative contaminant, also exhibits mass loss, as illustrated in Figure 1. The degree of mass loss observed for benzoate is similar to that observed for toluene (see Figure 1). Similar results were observed for salicylate and naphthalene. This indicates that the biodegradation behavior of the biotracers is at least generally representative of the biodegradation behavior of the contaminants. However, it is not possible with the currently available data to determine if this would hold true at the field scale.

CONCLUSIONS

The results discussed herein indicate that it is possible to characterize the general degradation potential for a selected site using model compounds (biotracers). This method may, therefore, be a useful addition to our arsenal of methods for evaluating the feasibility and performance of in-situ bioremediation. It is envisioned that the biotracer test would be used in conjunction with other methods to help enhance our understanding of the target system. Additional research is needed to examine the specific representativeness of selected biotracers for contaminants of interest, and to test the effectiveness of the biotracer method under various conditions.

Acknowledgements

This research was supported by grants from the U.S. Environmental Protection Agency and the National Institute of Environmental Health Science Superfund Basic Science Research Program. The assistance of Brent Cain, Bill Blanford, John McCray, Ken Bryan, and other University of Arizona students during the field experiment is greatly appreciated, as is the assistance of Gwynn Johnson for the ethanol analyses.

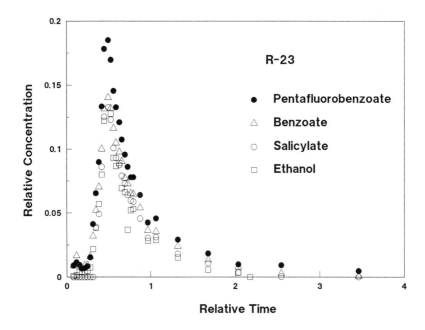

Figure 2.   Breakthrough curves measured at a multi-level sampling location (R-23) during the field experiment.

250

REFERENCES

1. Madsen, E.L. Environ. Sci. Technol., 25(10), 1663-1673, 1991.
2. NRC-National Research Council, In situ bioremediation: When does it work? National Academy Press, Washington, DC, 1993.
3. McAllister, P.M. and Chiang, C.Y. Ground Water Monitor. Remed., 14(2), 161-173, 1994.
4. Brockman, F.J. pp. 39-47 in: Monitoring and Verification of Bioremediation, Hinchee, R.E., Douglas, G.S., and Ong, S.K., eds., Battelle Press, Columbus, OH, 1995.
5. Gillham, R.W., Starr, R.C., and Miller, D.J. Ground Water, 28(6), 858-862, 1990.
6. Nielsen, P.H., Bjerg, P.L., Nielsen, P., Smith, P., and Christensen, T.H. Environ. Sci. Technol., 30(1), 31-37, 1996.
7. Barker, J.F., Patrick, G.C., and Major, D. Ground Water Monitor. Rev., 7, 64-71, 1987.
8. Semprini, L., Roberts, P.V., Hopkins, G.D., and McCarty, P.L. Ground Water, 28(5), 715-727, 1990.
9. MacIntyre, W.G., Boggs, M., Antworth, C.P., and Stauffer, T.B. Water Resour. Res., 20(12), 4045-4051, 1993.
10. Kuhn, E.P., Zeyer, J., Eicher, P., and Schwarzenbach, R.P. App. Environ. Micro., 54(2), 490-496, 1988.
11. Chapelle, F.H., **Ground-Water Microbiology & Geochemistry**, John Wiley and Sons, New York, NY, 1993.
12. Grbic-Galic, D., Chap. 3 in: **Soil Biochemistry**, J. Bollag and G. Stotzky, eds., Marcel Dekker, Inc, New York, NY, 1990.
13. Gelhar, L.W., Welty, C., and Rehfeldt, K.R. Water Resour. Res., 28(7), 1955-1974, 1992.
14. Cerniglia, C.E., Heitkamp, M.A., Chap. 2 in: **Metabolism of Polycyclic Aromatic Hydrocarbons in the Aquatic Environment**, U. Varanasi, ed., CRC Press, Boca Raton, FL, 1989.

Chapter 18

A Diffusive Tracer-Test Method for Investigating the Influence of Mass Transfer Processes on Field-Scale Solute Transport

Mark L. Brusseau[1,2], Q. Hu[1], N. T. Nelson[2], and R. Brent Cain[2]

[1]Departments of Soil, Water, and Environmental Science and [2]Hydrology and Water Resources, University of Arizona, Tucson, AZ 85721

The purpose of this work is to illustrate the utility of the diffusive tracer-test method for investigating the influence of diffusion-mediated processes on solute transport at the field scale. The diffusive tracer test involves the use of a suite of tracers of various sizes (i.e., diffusivities), and is based on the hypothsis that diffusion-mediated processes will generally be sensitive to the aqueous diffusivity of the solute. Hence, the transport of tracers with different diffusivities should be different when diffusion-mediated processes are important. The method was successfully applied to a field site representative of alluvial aquifer systems.

The relative importance of diffusion-mediated processes for solute transport at the field scale remains a subject of debate. For example, local-scale dispersion (including molecular diffusion) is usually ignored in field-scale modeling. It is also usually ignored in the development of stochastic-analytical theories of solute transport. However, it is recognized that this approach may not always be accurate. For example, it is accepted that vertical "mixing" can play an important role in the field-scale evolution of solute plumes. However, the potential contribution of diffusion-mediated processes to vertical mixing remains unclear. Based on the results of field experiments, several authors have speculated that diffusion-based processes contributed to observed solute transport (see review by Brusseau, 1). However, the impact of diffusion-mediated processes was not definitively characterized. Given the above, a method designed to investigate the contribution of diffusion-mediated processes to field-scale solute transport would appear to be useful.

Diffusion-mediated processes include diffusion coincident with the axes of flow (e.g., longitudinal diffusion), diffusion transverse to the flow axes, mass transfer across films of water coating soil particles, mass transfer between flowing and nonflowing water (dead-end porosity, intraparticle porosity), and mass transfer associated with heterogeneity (aggregated, layered, and fractured media). Methods for determining the

the laboratory scale, the primary method used to date involves conducting tracer experiments at two or more pore-water velocities. Because diffusion is a time-dependent process, changes in residence time (pore-water velocity) will influence the impact of a diffusion-mediated process on solute transport. This is illustrated in Figure 1a, where the breakthrough curve measured at a larger velocity exhibits greater asymmetry and tailing compared to one measured at a lower velocity for transport of a nonreactive tracer in a heterogeneous, aggregated porous medium. In this system, transport is influenced by diffusive mass transfer of solute between advective (interaggregate porosity) and nonadvective (intraaggregate porosity) regions. The shorter residence time associated with the larger velocity provides less time for mass transfer, which results in a greater degree of nonideal transport. Conversely, pore-water velocity has no influence on the breakthrough curves obtained for transport of a nonreactive tracer through a homogeneous porous medium where advection dominates transport (see Figure 1b).

While successful in the laboratory, the use of residence-time perturbations at the field scale may be problematic. First, the costs associated with conducting multiple field experiments (multiple velocities) are often prohibitive. Second, the resolution associated with field experiments would normally require at least an order-of-magnitude difference in pore-water velocities to ensure sensitivity to diffusion-mediated processes. The time associated with conducting lower velocity experiments may often be prohibitive. The residence-time perturbation method can be implemented during a single test through the use of multiple sampling points, located at different distances from the injection point. However, the additional requirements associated with this approach may often preclude its application.

Another method that has been used to examine the influence of rate-controlled processes on solute transport at the laboratory scale is the "flow-interruption" method (4). A temporary cessation of flow during the transport of a solute influenced by a rate-controlled process will generally cause a perturbation in the concentration profile. This method has been used to examine the impact of diffusion-mediated processes on the transport of solute in locally heterogeneous porous media (3,5,6). However, as discussed by Brusseau et al. (7), the application of this method to the field may be constrained by the relative time scales of advection and diffusion.

A third method, the diffusive tracer test, presented by Brusseau (2), involves the use of a suite of tracers of various sizes (i.e., diffusivities). Clearly, a diffusion-mediated process will generally be sensitive to the aqueous diffusivity of the solute. Hence, the transport of tracers with different diffusivities should be different when diffusion-mediated processes are important. This is illustrated in Figure 2a, which shows that the breakthrough curve for a larger nonreactive solute exhibits greater asymmetry and tailing compared to that of a smaller nonreactive solute for transport in an aggregated soil. Conversely, solute diffusivity has no influence on the breakthrough curves obtained for transport of nonreactive tracers through a homogeneous porous medium where advection dominates transport (see Figure 2b).

The comparison presented in Figure 2 illustrates the influence of solute diffusivity on transport for cases where solute retention is influenced by rate-limited mass transfer (e.g., diffusion into intraaggregate or intraparticle domains). The opposite

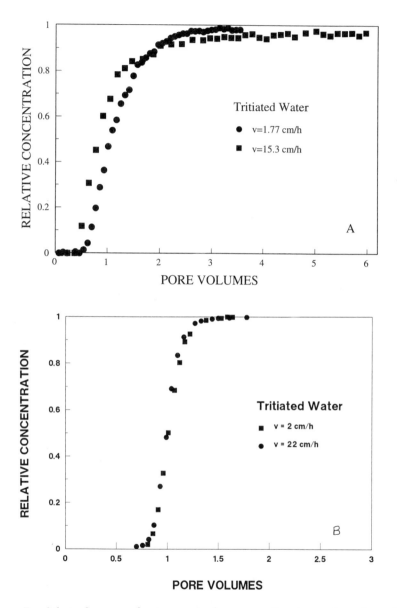

1.   Breakthrough curves for transport of a nonreactive tracer through packed
     columns: A) heterogeneous, aggregated medium (from 3), B) homogeneous soil
     (from 2).

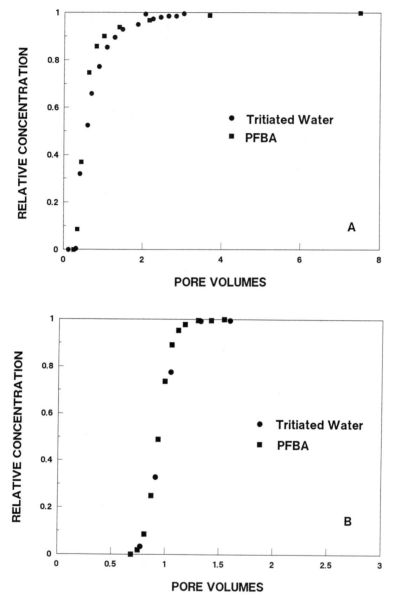

2.   Impact of solute diffusivity on transport: A) heterogeneous, aggregated soil (v = 70 cm/h; from 2), B) homogeneous soil (v = 88 cm/h; from 2), C) homogeneous medium (v = 0.06 cm/h; from 8).

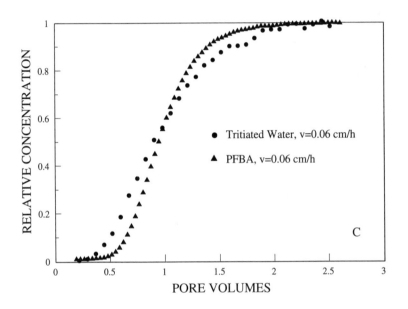

Figure 2. *Continued.*

effect is observed for systems wherein axial diffusion is important. For example, under conditions where longitudinal diffusion is important (low pore-water velocities), the breakthrough curve for a larger solute exhibits less spreading compared to that of a smaller solute for transport in a homogeneous medium (see Figure 2c). Conversely, as shown previously, solute diffusivity has no influence on the breakthrough curves obtained for transport of nonreactive tracers through a homogeneous porous medium when advection dominates transport (Figure 2b).

The purpose of this work is to illustrate the utility of the diffusive tracer-test method for investigating the influence of diffusion-mediated processes on solute transport at the field scale. The method is applied to a field site representative of alluvial aquifer systems. A dual-porosity transport model is applied to the results of the tracer tests to help evaluate the impact of solute diffusivity on transport.

MATERIALS AND METHODS

Tracers

Two tracer tests were conducted, and a different set of tracer pairs was used for each test. Bromide was used as the typical nonreactive tracer. Pentafluorobenzoate was used as a nonreactive tracer whose size and diffusivity is representative of many organic contaminants of environmental concern. Hydroxypropyl-beta-cyclodextrin (HPCD) and sodium mono-and-dimethylnaphthalene sulphonate (SMDNS) were used as the larger-sized, smaller-diffusivity tracers. HPCD is a glucose-based molecule that has a polar exterior surface, and is about 1 nm in diameter. It has been used in a number of laboratory studies, the results of which indicate that HPCD has little reactivity with soil (negligible sorption) and does not appear to experience pore-exclusion phenomena (3,9). SMDNS is a cosurfactant and also experiences minimal sorption (10). Pertinent properties of the tracers are listed in Table 1. Inspection of Table 1 reveals that the diffusion coefficient for SMDNS is 4 times larger than that of bromide, and that the diffusion coefficient for PFBA is 4.4 times larger than that of HPCD.

Field Site

The field site is located at Hill Air Force Base in Layton, UT. The two tracer tests were conducted in a portion of an unconfined aquifer that consists of fine to coarse sand, interbedded with gravel and silt layers. The formation is a deltaic deposit, with braided channels and large numbers of cobbles. The natural saturated thickness of this unit ranges from zero to 3 meters. A thick clay unit, the top of which is about 10 meters below ground surface, underlies the aquifer. The average porosity was estimated to be 0.18, and the mean horizontal hydraulic conductivity was calculated to be $4.5 * 10^{-2}$ cm/s, both based on the results of the tracer tests. The tracer tests were conducted in enclosed 3×5 m^2 cells emplaced in the subsurface. The cells were created by driving sealable-joint sheet-pile walls into the clay layer. The cells were created with minimal disruption of the subsurface located within the cells. A line of four fully screened injection wells and a line of three fully screened extraction wells, both normal to the direction of flow and located 4.5 meters apart, were used to create a steady state flow field. This design allows solute transport to be modeled as one-dimensional at any

Table 1. Tracer Properties.

Tracer	Molecular Weight	D_0 (cm²/h)	C_0 (mg/L)
Tracer Test #1			
Bromide	79.9	0.075	385
SMDNS	260	0.0187	200
Tracer Test #2			
PFBA	212	0.0284	262
HPCD	1500	0.0064	281

single sampling location, especially in the center of the system where boundary effects are minimal.

After steady state flow was established, a 0.2 pore-volume pulse of tracer solution was injected into the cell, followed by approximately ten pore volumes of tracer-free water [pore volume defined by travel time of PFBA]. The influent concentrations for the four tracers are listed in Table 1. A total flow rate of approximately one pore volume per day was maintained throughout the experiment (Q ≈ 4.5 l/min). Flow was monitored and adjusted to maintain the water table at 5.4 ± 0.15 meters below ground surface, giving a saturated thickness of 3 meters.

A three-dimensional, multi-level sampling array (12 locations, 5 depth intervals) connected to a vacuum system was used to collect depth-specific water samples. Depth-integrated samples were collected from the three extraction wells. Samples were collected with minimal headspace in polyethylene vials, and stored at 4 °C until analysis. Bromide was analyzed using a colorimetric method (Alpkem, Clackamas, OR). Pentafluorobenzoate and SMDNS were analyzed by UV-VIS high performance liquid chromatography (Waters). HPCD was analyzed using a fluorescence-based method (Hitachi Fluorescence Spectrophotometer) (11).

Mathematical Modeling

It is known that the subsurface at the field site is heterogeneous. However, the exact nature of the heterogeneity, such as the spatial variability of hydraulic conductivity, is unknown. This makes the application of stochastic-based transport models problematic for this site. An alternative, simpler approach is to use the dual-porosity model to approximate the influence of physical heterogeneity on solute transport. The dual-porosity model has been used successfully to simulate solute transport in heterogeneous aquifers at small field scales (12-14).

The nondimensional governing equations for one-dimensional transport assuming steady state flow are:

$$\beta R \frac{\partial C_a^*}{\partial T} + (1-\beta)R \frac{\partial C_n^*}{\partial T} = \frac{1}{P}\frac{\partial^2 C_a^*}{\partial X^2} - \frac{\partial C_a^*}{\partial X} \tag{1}$$

$$(1-\beta)R \frac{\partial C_n^*}{\partial T} = \omega(C_a^* - C_n^*) \tag{2}$$

where the following nondimensional parameters are defined as:

$C_a^* = C_a/C_0$	(3a)	$C_n^* = C_n/C_0$	(3b)
$X = x/l$	(3c)	$T = tv/l$	(3d)
$P = ql/D\theta_a$	(3e)	$R = 1 + \rho K_d/\theta$	(3f)
$\beta = (\theta_a + f\rho K_d)/(\theta + \rho K_d)$	(3g)	$\omega = \alpha l/v\theta_a$	(3h)

and where C_a is concentration of solute in the advective solution (M L^{-3}), C_n is solute concentration in the nonadvective domain (M L^{-3}), C_0 is the concentration of solute in the influent solution, t is time (T), x is distance (L), l is characteristic length (L), ρ is bulk density of the porous medium (M L^{-3}), n is the porosity, θ_a is the fraction of porosity in which advection occurs, q is Darcy flux (L T^{-1}), v is average pore-water velocity (L T^{-1}), D is the hydrodynamic dispersion coefficient (L^2 T^{-1}), K_d is the equilibrium sorption coefficient (L^3 M^{-1}), f is the fraction of sorbent associated with the advective domain, and α is the first-order mass transfer coefficient (T^{-1}).

The governing equations are solved with a finite-difference numerical approach under the following initial and flux-type boundary conditions:

$$C_a^*(X,0) = C_n^*(X,0) = 0 \qquad (4a)$$

$$C_0^* = C_a^* - \frac{1}{P} \frac{\partial C_a^*}{\partial X}\Big|_{x=0} \qquad (4b)$$

$$\frac{\partial C^*(1,T)}{\partial X} = 0 \qquad (4c)$$

The dual-porosity model is designed to allow the spreading and tailing associated with advection (differential rates of flow in heterogeneous media, local hydrodynamic dispersion) to be separated from the spreading and tailing associated with rate-limited diffusive mass transfer between zones that experience significant and minimal advective flux, respectively. The Peclet Number (P) represents the advection related spreading, whereas the impact of inter-region diffusive mass transfer on transport is represented by the ω parameter. This parameter is a ratio of the effective hydraulic residence time and the characteristic time of mass transfer. The larger the magnitude of ω, the closer the system is to a condition of equilibrium between the advective and nonadvective regions, and the smaller is the impact of inter-region mass transfer on transport (e.g. less tailing).

The first-order mass transfer coefficient, α, can be related to diffusion parameters as follows:

$$\alpha = \frac{a D_0 \theta_n}{\tau l^2} \qquad (5)$$

where D_0 is the aqueous diffusion coefficient (L^2 T^{-1}), "a" is a factor representing the shape of the nonadvective domains, τ is the tortuosity factor, and "l" is a characteristic length of the nonadvective domains. The impact of solute diffusivity on the magnitude of ω is clear by substituting equation (5) into equation (3h). Thus, we would expect a solute with a smaller D_0 to exhibit a greater degree of nonideal transport (i.e., smaller ω) compared to a solute with a larger D_0 (larger ω) for a given system.

RESULTS AND DISCUSSION
Extraction Well versus Multi-level Sampling Points

The breakthrough curves obtained for PFBA and HPCD at a specific multi-level sampling location (R23) and at the center extraction well (E52) are shown in Figure 3 and Figure 4, respectively. Both tracers exhibit a fair amount of tailing, due to the impact of heterogeneity on transport. The zeroth moment for HPCD is larger than that for PFBA because the input pulse for HPCD was larger than that for PFBA. It is also observed that the retardation factor for HPCD is less than 1. The possible reasons for this are not clear.

The measured breakthrough curves were well simulated by the dual porosity model (see Figures 3 and 4). The parameter values obtained by nonlinear least-squares optimization are reported in Table 2. Inspection of the Peclet Numbers, which represent the spreading associated with advection, reveals that the values obtained for the breakthrough curves measured at the multi-level sampling point are larger than those obtained for the extraction-well breakthrough curves. This is to be expected given that the extraction well is fully screened, and thus provides integrated sampling across the entire aquifer thickness.

The magnitude of the ω parameter, which represents the impact of inter-region (advective-nonadvective) mass transfer on transport, is larger for the breakthrough curves measured at the multi-level sampling points. For example, the mean of the values obtained for PFBA and HPCD breakthrough curves measured at R23 is 1.5, compared to a mean value of 0.27 for the breakthrough curves of those two tracers measured at E52. This behavior may again reflect differences in sampling domains associated with the extraction well and the multi-level system. Sampling across the entire aquifer thickness most likely provides greater opportunity to encounter zones that may effectively behave as nonadvective regions.

The Impact of Solute Diffusivity

Inspection of Table 2 shows that there is no statistically significant difference between the values of P, β, and ω determined for PFBA and HPCD at R23, the multi-level sampling point. Similar results were observed for bromide and SMDNS, the tracer pair used in the first test (Table 2). The similarity of the ω values indicates that there was no impact of solute diffusivity on transport for the region sampled between the injection and multi-level monitoring point. This may indicate that transport was not sensitive to the diffusivity difference between PFBA and HPCD or between bromide and SMDNS. It is possible that differences in transport (ω values) would have been observed if the difference in D_0 values would have been greater.

In contrast to the results obtained for the multi-level sampling points, differences in the magnitudes of ω were observed for the data measured at the extraction well. Specifically, the ω value determined for HPCD (0.09) was smaller than that determined for PFBA (0.44). This indicates that HPCD experienced greater nonideal transport compared to PFBA. As discussed above, differences in ω values correspond directly to differences in D_0 values for a given system. The difference in D_0 values between PFBA and HPCD is 4.4 (Table 1). This compares very well to the difference of 4.9 in the ω values determined for the two tracers. Thus, the expected difference in the impact

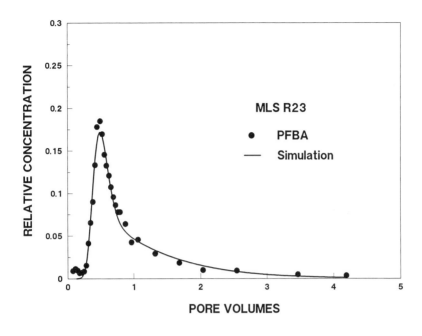

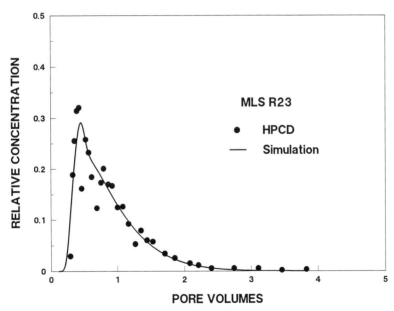

3. Breakthrough curves measured at multi-level sampling point (R23): A) pentafluorobenzoate (PFBA), B) cyclodextrin (HPCD).

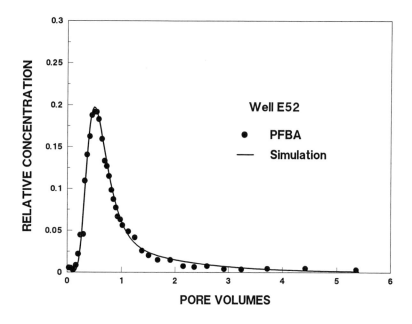

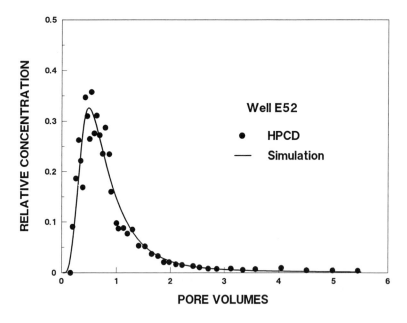

4.   Breakthrough curves measured at the center extraction well (E52): A) pentafluorobenzoate (PFBA), B) cyclodextrin (HPCD).

Table 2. Parameter Values Obtained from Optimization with Dual Porosity Transport Model.

Sample Point	R	P	β	ω
Tracer Test #1				
Bromide, R24	1.0	>100	0.57 (0.52-0.63)	5.4 (4.1-6.7)
SMDNS, R24	1.0	>100	0.47 (0.32-0.63)	11.5 (4.4-18.5)
Tracer Test #2				
PFBA, R23	1.0	29.3 (23.4-35.3)	0.48 (0.46-0.49)	1.0 (0.89-1.14)
HPCD, R23	0.8	67.4 (20-107)	0.39 (0.24-0.53)	2.0 (1.10-2.90)
PFBA, E52	1.0	8.2 (6.9-9.5)	0.59 (0.56-0.62)	0.44 (0.34-0.54)
HPCD, E52	0.9	2.5 (1.5-3.5)	0.80 (0.70-0.90)	0.09 (0.03-0.15)

Values in parentheses represent 95% confidence intervals.

of inter-region mass transfer on transport of the two tracers, based on their differences in diffusivity, matches with the measured behavior.

SUMMARY

The results of the experiment presented above illustrate that the diffusive tracer-test method presented by Brusseau (2) can be used to investigate the impact of diffusion-mediated processes on solute transport. A key to the performance of the method is selecting a tracer pair that will provide sufficient sensitivity (i.e., sufficient difference in D_0 values) for the specific system. The diffusive tracer test can be combined with other techniques, such as flow perturbation, to enhance sensitivity. The test case presented herein represented an alluvial aquifer system. Another, larger-scale diffusive tracer test was conducted recently in another alluvial system, and similar results were observed (15). It is expected that the diffusive-tracer method may be especially useful for systems composed of fractured or highly aggregated media.

Acknowledgements

We wish to thank Carl Enfield, Lynn Wood, and several University of Arizona students for their assistance during the experiments. We also thank Joe Piatt, Xiaojiang Wang, and Wouter Noordman for their assistance in tracer analysis. This research was supported by grants provided by the U.S. Environmental Protection Agency and the National Institute of Environmental Health Sciences (Superfund Basic Research Program).

REFERENCES

1. Brusseau, M.L. Rev. Geophysics, 1994, 32, 285-313.
2. Brusseau, M.L. Water Resour. Res., 1993, 29, 1071-1080.
3. Hu, Q.; Brusseau, M. L. Water Resour. Res., 1995, 31, 1637-1646.
4. Brusseau, M.L.; Rao, P.S.C.; Jessup, R.E.; Davidson, J.M. J. Contam. Hydrol., 1989, 4, 223-240.
5. Koch, S.; Fluhler, H. J. Contam. Hydrol., 1993, 14, 39-54.
6. Hu, Q.; Brusseau, M. L. J. Contam. Hydrol., 1996, 24, 53-73.
7. Brusseau, M.L.; Hu, Q., Srivastava, R. J. Contam. Hydrol., 1997, 24, 205-219.
8. Hu, Q.; Brusseau, M. L. J. Hydrol., 1994, 158, 305-317.
9. Brusseau, M.L.; Wang, X.; Hu, Q. Environ. Sci. Technol., 1994, 28, 952-956.
10. Shiau, B.; Sabatini, D.A.; Harwell, J.H. Environ. Sci. Technol., 1995, 24, 2929-2235.
11. Wang, X; Brusseau, M.L. Environ. Sci. Technol., 1993, 27, 2821.
12. Goltz, M.N.; Roberts, P.V. J. Contam. Hydrol., 1986, 1, 77-93.
13. Goltz, M.N.; Roberts, P.V. J. Contam. Hydrol., 1988, 3, 37-63.
14. Brusseau, M.L. Water Resour. Res., 1992, 28, 2485-2497.
15. Nelson, N.T.; Hu, Q.; Brusseau, M.L. 1997 (in review).

Chapter 19

Contaminant Transport and Fate in a Source Zone of a Chlorinated-Solvent Contaminated Superfund Site: Overview and Initial Results of an Advanced Site Characterization Project

Mark L. Brusseau[1,2], J. W. Rohrer[2], T. M. Decker[1], N. T. Nelson[2], and W. R. Linderfelt[1]

[1]Departments of Soil, Water, and Environmental Science and [2]Hydrology and Water Resources, University of Arizona, Tucson, AZ 85721

An advanced site characterization project is being conducted at a chlorinated-solvent contaminated Superfund site to help improve the effectiveness of the remediation program. As a part of this project, two forced-gradient tracer tests were conducted to characterize the transport behavior of bromide and resident trichloroethene and dichloroethene in the aquifer underlying a contaminant source zone. The results indicate significant vertical spatial variability of hydraulic (hydraulic conductivity) and chemical (contaminant concentrations) variables over the 6 m thick aquifer zone. The tracer experiments provide a direct demonstration of the influence of subsurface heterogeneity on water flow, and its possible impact on the efficient removal of trichloroethene and dichloroethene by pump and treat. Extensive elution tailing and rebound were observed for trichloroethene and dichloroethene during the experiments, which indicate the presence of a significant mass of contaminant whose transfer to the advecting water is rate limited. This mass could be sorbed to aquifer material, located within low permeability zones, associated with an immiscible-liquid phase, or some combination thereof. Identification of the factors that control contaminant mass removal is critical for a complete, successful cleanup of the site. Current and future research associated with this site targets these issues.

The fate of contaminants associated with hazardous waste sites, landfills, and other sites of chemical use and disposal has become a predominant environmental issue. Enormous amounts of time and resources are being expended on identifying potential problem sites, conducting risk assessments, and on planning and conducting remediation programs. Success in these endeavors requires an understanding of the transport and fate of contaminants at the field scale. Unfortunately, the scope of current characterization programs are usually insufficient to provide a clear understanding of contaminant transport for the site of concern. However, the use of innovative site characterization approaches, coupled with advances in sampling and monitoring technologies, has provided a means to obtain the information necessary to develop a better understanding of contaminant transport and fate. This paper provides an overview and the initial results of an advanced characterization study conducted at a Superfund site in Tucson, Arizona.

Background

The primary objectives of our project are to: (1) characterize the transport and fate behavior of resident contaminants at the Superfund site, (2) identify the factors controlling the effectiveness of the currently operating pump-and-treat system, and (3) provide information useful for improving the remediation program. The project is based on an integrated approach that consists of several components, including preliminary site-characterization activities, laboratory experiments, field tracer tests, and mathematical modeling. The specific components of the project are listed in Table I. This paper will focus on the results of the first two field tracer experiments. A brief background for the site is presented below.

Site History. The site at which the experiments were conducted is part of the Tucson International Airport Area Superfund site, in Arizona. The site was placed on the National Priorities list in August 1983 in response to the detection of trichloroethene in several wells (1). A large, multiple-source plume of trichloroethene and 1,1-dichloroethene exists in the upper portion of the regional aquifer (see Figure 1), which is the sole source of potable water for the city. Contaminants entered the subsurface by seepage from pits and ponds used to dispose of organic solvents, including trichloroethene, during the late 1950's to mid 1970's. The experiment site coincides with the former location of a large, unlined disposal pit (18.3 m x 3.7 m deep x 6.1 m wide), which is considered a contaminant source zone. In 1976 the pit was closed, and a percolation pond was built on top of the pit and used for several years. Various remediation programs have been in operation at different parts of the Superfund site for several years.

In 1987 a large pump-and-treat groundwater remediation project was initiated in the vicinity of the experiment site, which in February 1996 included 24 extraction wells and 20 recharge wells (2). The pump-and-treat system has been effective at containing the plume and decreasing its size, as shown in Figure 1. As of February 1996, approximately 8,000 Kg of trichloroethene had been removed from the groundwater (2) and the size of the plume defined by trichloroethene concentrations greater than 100 $\mu g/L$ had been decreased in area by 96% (3). However, the system has begun to exhibit reduced efficiency, as illustrated by the marked tailing observed for the trichloroethene concentration in the composite extracted groundwater entering the treatment plant (see Figure 2). In addition, concentration rebound has been observed for extraction wells located within the source zones.

Geology and Hydrogeology. The site is located in the Tucson Basin, which is underlain by several thousand feet of alluvial sediments interbedded locally with volcanic flows, agglomerates, and tuffaceous sediments. The major hydrogeologic units in the area of the site have been designated as (in descending order with depth):

Table I. Components of the Advanced Site Characterization Project.

Component	Purpose
Field	
Initial Hydraulic Characterization	Determine gradient, transmissivity
Nonreactive Tracer Test	Determine velocities, dispersivities
Partitioning Tracer Test	Detect and quantify immiscible liquid saturation
Diffusive Tracer Test	Characterize impact of diffusive mass transfer
Vertical Sampling (cores and aqueous)	Characterize vertical contaminant distribution
Contaminant Elution and Rebound	Characterize contaminant removal behavior
Laboratory	
Sorption Experiments[1]	Measure magnitude and linearity of sorption
Sorption Experiments[2]	Measure desorption of TCE from aquifer solids
Undisturbed Core Expts	Examine impact of core-scale heterogeneity
Intermediate-Scale Expts	Examine impact of interm.-scale heterogeneity
Modeling	
Experiment-Scale Modeling	Characterize transport at experiment scale
Plume-Scale Modeling--Flow	Characterize flow field at plume scale
Plume-Scale Modeling--Transport	Characterize TCE transport at plume scale
Remediation Modeling	Simulate operation of pump-and-treat system

Experiments conducted using [1]uncontaminated or [2]contaminated aquifer material.

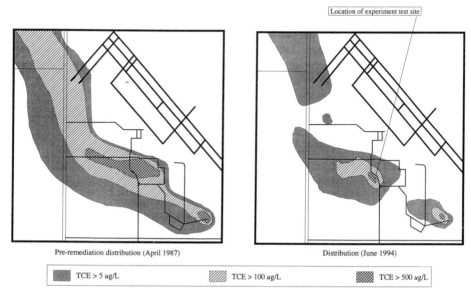

Pre-remediation distribution (April 1987)   Distribution (June 1994)

Location of experiment test site

TCE > 5 ug/L   TCE > 100 ug/L   TCE > 500 ug/L

Figure 1. Concentration of trichloroethene in groundwater in the vicinity of the tracer experiment site.

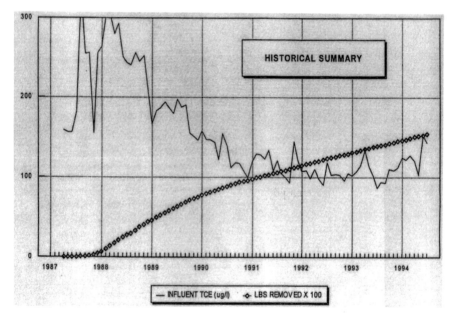

HISTORICAL SUMMARY

— INFLUENT TCE (ug/l)   ◇ LBS REMOVED X 100

Figure 2. Concentration of trichloroethene entering the treatment plant; the concentration represents a composite of all extraction wells.

the Unsaturated Zone, the Upper Aquifer, an Aquitard Unit and the Lower Aquifer (*4,5*). The Unsaturated Zone extends from the land surface to the regional groundwater surface, generally located at 30 to 40 m below ground surface (BGS). The Upper Aquifer extends from the regional groundwater surface to a depth of approximately 54 to 66 m BGS, where it is underlain by the Aquitard Unit. The Upper Aquifer is comprised of sand and gravel lenses/layers separated in some areas by clayey sediment ranging from one meter to greater than seven meters in thickness. In the vicinity of the source zone wherein the tracer experiments were conducted, the aquifer is divided in two by a relatively continuous low permeability layer. The field tracer experiments were conducted in the coarser grained layer extending from approximately 43-49 m BGS in the upper portion of the Upper Aquifer. The aquifer in this area consists of an approximately 6 m thick zone of higher permeability clayey sands and gravels bounded above and below by lower permeability units. The higher permeability unit appears laterally extensive directly under the experiment site, as evidenced by the presence of a coarser grained sequence at approximately 42-49 m BGS in all local wells.

The aquifer fines upwards from a clayey gravel to a clayey medium sand and is indicative of a typical sequence of arid valley alluvial deposits. Above the aquifer is a clay to sandy clay, ranging from 1.5 to 7.5 m in thickness, that appears to act as a semi-confining unit. Below the permeable unit is a high plasticity clay. The results of a sieve analysis of cuttings recovered from 43-45 m BGS during the drilling of well M-72 showed a composition of 26.4% gravel, 65.9% sand, 3.0% silt, and 4.7% clay. An analysis of the less than 2-mm fraction showed 0.06% total carbon and 0.03% organic carbon.

The potentiometric surface in the study area is approximately 37 m BGS. Transmissivity in the Uppermost Aquifer underlying the site has been estimated to be 100.6 m^2/d (*4*). As part of the initial site characterization, we conducted several aquifer tests to better define the site-scale hydrogeology. Analysis of this work produced a transmissivity for the Uppermost Aquifer of 55.9 m^2/d, a storativity of 0.0018 and, given a 6 m thick section, an average hydraulic conductivity of about 9 m/day (*6*).

Groundwater Chemistry and Contamination. Calcium and sodium are the dominant cations and bicarbonate and sulfate represent the dominant anions in the Upper Aquifer. The total dissolved solids (TDS) ranges from 230-500 mg/L. Trichloroethene and, to a lesser extent, 1,1-dichloroethene are the primary organic contaminants at the site. The contaminants are limited to the Upper Aquifer, especially the upper portion. Concentrations of trichloroethene in the groundwater range from less than 100 to about 10,000 μg/L in the vicinity of the experiment site.

Materials and Methods

Two dual-well, forced-gradient tracer experiments were conducted at the site in the summer and fall of 1995. The injection (SVE-6) and extraction wells (SVE-7) were 7.5 m apart (see Figure 3). Bromide was used as a nonreactive tracer to

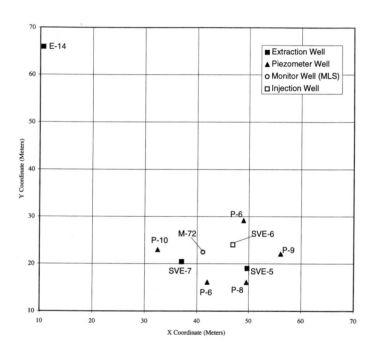

Figure 3. Location of pertinent wells.

examine flow variability, travel times, and contaminant fate. The elution of trichloroethene and dichloroethene was also examined.

A schematic of the injection apparatus is shown in Figure 4. Two days before experiment startup, the tracer-stock tank was filled with about 3800 liters of water. Calcium bromide ($CaBr_2$) was then added to the tank. Tracer-free water was used during the injection to dilute the stock solution at a ratio of about 100:1. The source of the tracer-free water was treated groundwater extracted from the site and groundwater from the city potable water supply for the first and second experiments, respectively. For the second experiment, a pressure regulator was installed in parallel with a valved pipe to aid in stabilizing flow of tracer-free water.

Concentrated tracer solution was injected into the tracer-free water line with a venturi injector device for the first experiment, and a peristaltic pump for the second experiment. The desired stock-solution flow rate was 1.1 liters/min (lpm), while the target tracer-free water flow rate was 112.4 lpm, which gave a total injection rate of 113.5 lpm for the first experiment. The injection rate for the second experiment was 136.3 lpm. The extraction rates were 113.5 and 136.3 lpm, for the first and second experiments, respectively.

A 1½ inch (internal diameter) PVC Schedule 40 pipe with a packer, connected to a 2 inch galvanized riser pipe, was used to introduce the tracer solution into the injection well at 43.5 m BGS. The packer was set 1.5 m above the end of the PVC pipe. For the second experiment, the end of the down-well portion of the pipe was capped and had two holes drilled in it to act as orifices. The orifices were used to maintain a positive pressure in the injection piping.

Sampling for bromide and trichloroethene/dichloroethene was conducted at a centerline monitor well (M-72), located 4.2 m from the injection well (SVE-6), at the extraction well (SVE-7), and at several monitoring wells located around the couplet (Figure 3). Samples were collected prior to initiation of the experiment to establish background concentrations for bromide and initial concentrations for trichloroethene and dichloroethene. Samples at well M-72 were collected using a multi-level gas drive sampler, described below. Extraction well samples were collected from a port on the discharge line, and samples for the peripheral monitor wells were collected by bailing.

A custom multi-level sampling device (Burge Environmental Inc., Tempe, AZ) was used to collect small, representative samples from vertically discrete zones (the locations of the ports are listed in Table II). The device has four sampling ports, with each port segregated within the screened well casing by a packer assembly above and below. The length of separation is adjustable from a minimum of 0.3 m. The packers are of a novel design, wherein gas pressure is used to decrease their cross-sectional area to allow emplacement of the sampling device. Once emplaced, the pressure is removed and the packers expand to full contact with the well casing. This design eliminates the potential of gas leaks associated with positive pressure gas-filled packers.

The device contains a single, glass sample-collection chamber (100 ml). A series of solenoid valves allows switching access to this chamber from one sample port to another. When a valve is activated, the chamber fills by hydrostatic pressure.

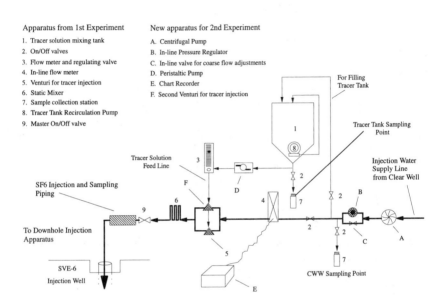

Apparatus from 1st Experiment

1. Tracer solution mixing tank
2. On/Off valves
3. Flow meter and regulating valve
4. In-line flow meter
5. Venturi for tracer injection
6. Static Mixer
7. Sample collection station
8. Tracer Tank Recirculation Pump
9. Master On/Off valve

New apparatus for 2nd Experiment

A. Centrifugal Pump
B. In-line Pressure Regulator
C. In-line valve for coarse flow adjustments
D. Peristaltic Pump
E. Chart Recorder
F. Second Venturi for tracer injection

For Filling
Tracer Tank

Tracer Tank Sampling
Point

Tracer Solution
Feed Line

Injection Water
Supply Line
from Clear Well

SF6 Injection and Sampling
Piping

To Downhole Injection
Apparatus

SVE-6
Injection Well

CWW Sampling Point

Figure 4. Schematic of tracer injection apparatus.

Table II. Moments, Pore-Water Velocities, Hydraulic Conductivities, and Dispersivities for Experiment 2.

Zone ID	Depth (M BGS)	M_0 (Day)	M_1 (Day^2)	T_t (Day)	Velocity (M/Day)	K (M/Day)	Dispersivity (M)
A2	42.7	1.8	10.8	5.1	0.8	0.9	N.D.
C2	44.2	2.2	3.4	0.5	8.3	9.5	0.5
D2	45.1	2.2	3.1	0.4	10.0	11.4	0.3
B2	47.6	2.1	35.6	16.0	0.3	0.3	N.D.

BGS = below ground surface; M_0 = zeroth moment; M_1 = first moment; T_t = mean travel time; N.D. = not determined because curves exhibited multiple peaks.

The valve is closed automatically by operation of a water sensor located near the top of the chamber. The size of the sample is dictated by the position of the water level sensor. Once the chamber is filled, an exit valve is switched open and a stream of nitrogen gas pushes the water sample to the surface. A small-diameter tube (1/8 inch) is used for surface delivery, which minimizes the gas-water interface available for volatilization.

Trichloroethene and dichloroethene samples were collected with no headspace in 20-mL volatile organic analysis (VOA) vials. Bromide samples were collected in 20-mL polyethylene vials. Samples for trichloroethene and dichloroethene analysis were taken to an on-site analytical laboratory, where they were quantified by gas chromatography (Hall detector). Samples for bromide analysis were taken to the University of Arizona, where they were analyzed by use of a colorimetric test (ALPKEM Corporation, Clackmas, Oregon).

The first field tracer experiment was designed to determine the physical flow parameters of the higher conductivity zone in the Uppermost Aquifer and to test all the components of the experimental system. The tracer experiment was conducted in May, 1995. The tracer solution was injected into well SVE-6 for 60 hours, and tracer-free water was injected for an additional 87.5 hours. The second tracer experiment was conducted for approximately 10 weeks between August and October of 1995. The tracer solution was injected into well SVE-6 for 49.5 hours, and tracer-free water was injected for an additional 58 days. The main objective of the second experiment was to observe resident contaminant elution behavior in response to pumping and gain insight into factors controlling contaminant removal. For both experiments, samples were collected from M-72 after pumping had stopped to observe possible rebounding of trichloroethene and dichloroethene.

Results and Discussion

Bromide Transport: Injection and Extraction Well Data. The injection and extraction flow rates exhibited very little variability during the experiments ($\pm 10\%$ and $\pm 3\%$ for the first and second experiments, respectively). The volumes of water pumped during the two experiments were 1,026 and 11,862 m^3 for the first and second experiments, respectively. Water levels stabilized relatively rapidly and were similar for both experiments.

The average bromide injection concentration was determined from samples obtained at the tracer mixture sampling port. The average injection concentration for bromide was 40.8 mg/L ($\pm 20\%$) for the first experiment and 55.6 mg/L ($\pm 6\%$) for the second experiment. The total mass of bromide injected during the two experiments was 14.8 and 22.6 Kg for the first and second experiments, respectively.

The breakthrough curves for bromide at the extraction well (SVE-7) are shown in Figure 5 for both experiments. Their general shapes are similar, and are typical of those observed for a "fully screened" pumping well. The mass of bromide extracted was calculated using an average extraction flowrate and integrating under the curves. The mass recovery for bromide was 39% and 72% for the first and second experiments, respectively. The greater recovery for the second experiment is related to two factors. First, a greater pumping rate was used for the second

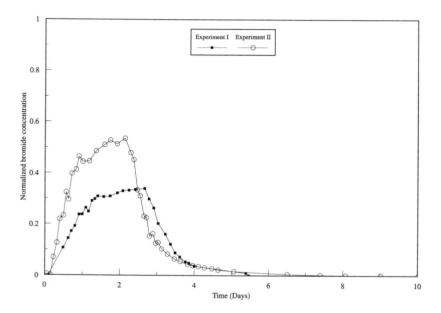

Figure 5. Breakthrough curves for bromide measured at the extraction well (SVE-7) for both experiments.

experiment, which should increase recovery in the presence of a background flow field. Second, a relatively large-volume extraction well (E-14) located 57 m from our couplet was pumping during the first experiment and not during the second. The mass loss observed for the second experiment is most likely the result of flow lines escaping from the injection-extraction well couplet due to the presence of areal heterogeneity.

Bromide Transport: Monitor Well Data. Breakthrough curves for bromide measured at levels C2 and D2 of the centerline monitor well were sharp and symmetrical, indicating a relatively small degree of spreading. Conversely, breakthrough curves for levels A2 and B2 exhibited extended tailing and lower peak concentrations. Moment analysis of these data was performed to evaluate mass recovery and travel time for each of the four monitoring points (see Table II). The zeroth moments for levels B2, C2, and D2 are very similar to the true injection pulse of 2.1 days, which indicates ideal behavior with respect to mass recovery and one-dimensional flow at these points. The mean travel times vary between 0.4 and 16 days, exhibiting the impact of heterogeneity on flow and advective transport.

Given that M-72 is located on the centerline of the couplet, pore-water velocities can be calculated from the travel times by using the straight-line distance from SVE-6 to M-72 (4.2 m). These values, which represent some mean value of the actual velocity profile between the injection and monitor wells, range from 0.3 to 10 m/d (Table II). Horizontal hydraulic conductivities can be estimated from these velocities if the hydraulic-head gradient and porosity are known. The average gradient between SVE-6 and M-72 was estimated to be 0.29 m/m, and a value of 0.33 was assumed for porosity (7). Using these values, the estimated horizontal hydraulic conductivities for levels A2-D2 are reported in Table II. These values are within the same range as values obtained from pump tests, which were mentioned previously.

The flow in the vicinity of a centerline monitoring well can usually be assumed to be one-dimensional (8), and the zeroth-moment data presented above indicate this is so for our experiments. Thus, the breakthrough curves measured for M-72 can be analyzed with a one-dimensional advection-dispersion model to determine longitudinal dispersivities. The program that was used to analyze the breakthrough curves was CXTFIT (9), which couples a nonlinear least-squares optimization routine to the advection-dispersion equation. For our analyses, the retardation factor was assumed to equal one, the size of the injection pulse was known from the experiment, and the dispersion coefficient and the pore-water velocity were fitted. The longitudinal dispersivity (α_L) was calculated from the fitted dispersion coefficient (D) assuming that diffusion was negligible compared to hydrodynamic dispersion [i.e., $D = \alpha_L v$]. Values for the dispersivity were about 0.4 m (Table II). These values are consistent with other values measured at similar scales (10). It must be noted that this approach assumes a constant velocity between the injection and monitor wells, which is not the actual case. Thus, the dispersivity values are minimum, "macrodispersivity" values. The fitted pore-water velocities matched those determined by moment analysis.

Trichloroethene and Dichloroethene Data. Smaller-scale vertically discrete concentration data are rarely collected at hazardous waste sites. However, it is important to delineate vertical concentration profiles to fully understand contaminant distributions and transport, and to ensure successful planning and operation of contaminant remediation systems. Several studies have shown that contaminant concentrations can change significantly over scales of several centimeters in the vertical direction. Thus, nested monitoring wells can not provide the needed resolution. The multi-level sampling device provides a greater resolution of the vertical distribution of trichloroethene concentrations within the aquifer.

The results of the measurements, taken at about every 0.4 m over the 6 m interval, are shown in Figure 6. The highest concentrations of trichloroethene (~2,500 μg/L) were found near the upper portion of the aquifer, which has a large fraction of clayey materials. Concentrations decrease gradually from ~2,500 μg/L at 41.1 m BGS to ~125 μg/L at 46.2 m BGS. This lower portion of the aquifer consists of sands and gravels. Slightly higher concentrations of trichloroethene (~450 μg/L) were obtained at 47.1 m BGS, which is coincident with a confining layer of cemented fine grained materials. These results clearly indicate the vertical variability of trichloroethene concentrations within the aquifer, and indicate a possible inverse correlation with porous-media texture.

Plots of trichloroethene and dichloroethene concentrations measured during the second tracer experiment are shown in Figure 7 for the extraction well and the centerline monitoring well. The concentrations decline rapidly at the beginning of the experiment. However, it is clear that extensive tailing is exhibited for the duration of the experiment. The degree of tailing is especially great for the extraction well. The concentration of trichloroethene remained at approximately 150 μg/L even though contaminant-free water was flushed through the aquifer for 60 days. This is equivalent to 71 pore volumes, given the flow rate and bromide travel time associated with the well. The extensive tailing indicates the presence of a significant mass of trichloroethene in the vicinity of the extraction well.

The last set of data points on the elution curves represent samples collected after the injection and extraction wells were turned off. It is evident that the concentration of trichloroethene and dichloroethene exhibit strong rebounding. The last data points indicate that the concentrations rebound to levels close to or higher than the initial concentrations. The increase in contaminant concentrations can be caused by any factor that limits the rate of mass transfer in the aquifer material. Rate-limited mass transfer can include desorption from aquifer materials, diffusion out of low permeability domains, and dissolution from immiscible-liquid phases.

Conclusions

The results of the tracer experiments indicate significant vertical spatial variability of hydraulic (hydraulic conductivity) and chemical (contaminant concentrations) variables within the contaminated aquifer system. The tracer experiments provided a direct demonstration of the influence of subsurface heterogeneity on water flow, and its possible impact on the efficient removal of trichloroethene and dichloroethene by pump and treat. The high-permeability gravel layers associated with the aquifer serve as major zones of preferential flow. In addition, these zones appear to be characterized by lower trichloroethene sorption.

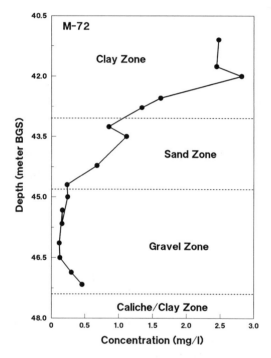

Figure 6. Vertical distribution of trichloroethene within the aquifer at well M-72.

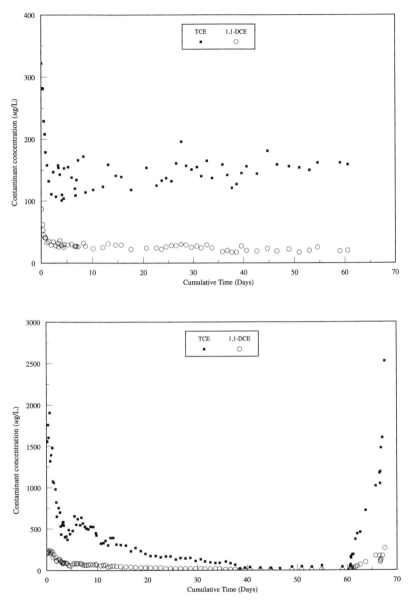

Figure 7. Concentrations of trichloroethene and dichloroethene measured during the second tracer experiment; A) extraction well, B) M-72, level B2.

These factors combine to reduce the difficulty of remediating these high permeability zones, as is indicated by their lower trichloroethene concentrations. However, high trichloroethene concentrations appear to be associated with the low permeability units. This, in combination with the low rates of flushing experienced by these zones, would increase the difficulty of using pump and treat to fully remediate the aquifer.

The strong elution tailing and rebound observed for trichloroethene and dichloroethene during the experiments indicates the presence of a significant mass of contaminant whose transfer to the advecting water is rate limited. This mass could be associated with the solid phase (sorbed), the low permeability zones discussed above, or with an immiscible liquid phase. Large amounts (about 2000 kg) of trichloroethene and other chlorinated solvents were removed during pilot-scale operations of soil venting in the vicinity of the tracer experiment site (3), which indicates the probable presence of immiscible-liquid phases in the vadose zone. It is uncertain, however, whether or not immiscible-liquid phases exist in the saturated zone. Identification of the specific factors that control contaminant mass removal is critical for a complete, successful site cleanup. Current and future research associated with this site targets these issues.

Acknowledgements

This research was supported by the United States Air Force/ASC/EMR Project F33657-81-E-2096. Several people were instrumental in the gathering of this data and deserve recognition: Bill Taylor of Taylor Controls who helped design the injection apparatus and helped out in many other ways; HMSC personnel who helped in the implementation of this project; Burge Environmental who developed the multilevel sampler and provided support; Denise Putz and other University of Arizona researchers who took time out of busy schedules to help in this project.

References

1.   U. S. Environmental Protection Agency, Superfund: progress at national priority list sites, Arizona and Nevada 1995 update, *Rep. EPA/540/R-95/073*, Environ. Prot. Agency, Washington, D. C., 1995.

2.   Hughes Missile Systems Company, United States Air Force Plant 44, Environmental Remediation Monthly Update - February 1996, Tucson, Arizona, March 14, 1996a.

3.   Hughes Missile Systems Company, United States Air Force Plant 44, Groundwater Remediation 1995 Annual Report, Tucson, Arizona, March 15, 1996b.

4.   Groundwater Resources Consultants Inc., Characterization of site hydrogeology in the vicinity of RCRA Impoundments, U. S. Air Force Plant No.44, Tucson, Arizona, Volume I, report *9106M-01*, February 6, 1991.

5.   Hargis and Montgomery Inc., Phase I investigation of subsurface conditions in the vicinity of abandoned waste disposal sites, Hughes Aircraft Company Manufacturing Facility, Tucson, Arizona, Volume I., 1982.

6.   Brusseau, M.L., J.W. Rohrer, W.R. Linderfelt, T.M. Decker, N.T. Nelson, Advanced characterization study to improve the efficiency of pump and treat operations at Air Force Plant 44: An integrated field, laboratory, and modeling approach, *Interim Progress Report*, August 18, 1995.

7.   Rohrer, J.W.   A field study of non-reactive transport behavior in a heterogeneous aquifer unit, M.S. Thesis, University of Arizona, 1996.

8.   Chrysikopoulos, C.V., P.V. Roberts, and P.K. Kitanidis, Water Resour. Res., 26(6), 1189-1195, 1990.

9.   Parker, J.C., and M.Th. van Genuchten, Determining transport parameters from laboratory and field tracer experiments, Virginia Agric. Experiment Station *Bull. 84-3*, 1984.

10.   Gelhar, L.W., C. Welty and K.R. Rehfeldt, Water Resour. Res., 28(7), 1955-1974, 1992.

INDEXES

Author Index

Subject Index

A

Acoustic enhanced remediation. *See* Sonic (acoustic) enhanced remediation

Advanced site characterization project
breakthrough curves for bromide at extraction well, 275*f*
bromide mass recovery, 274, 276
bromide transport-injection and extraction well data, 274–276
bromide transport-monitor well data, 276
components of project, 267*t*
concentration of trichloroethene entering treatment plant and in groundwater in vicinity of tracer experiment site, 268*f*
concentrations of trichloroethene and dichloroethene during second experiment, 277, 279*f*
custom multi-level sampling device, 271, 274
fate of contaminants associated with sites of hazardous waste disposal, 265–266
field components of project, 267*t*
flow in vicinity of centerline monitoring well, 276
geology and hydrogeology of site, 266, 269
groundwater chemistry and contamination, 269
history of Superfund site in Arizona, 266
laboratory components of project, 267*t*
location of injection and extraction wells, 270*f*
materials and methods, 269–274
modeling components of project, 267*t*
moments, pore-water velocities, hydraulic conductivities, and dispersivities for second experiment, 273*t*
primary objectives, 266

results from multi-level sampling device, 276–277
sampling method for bromide and trichloroethene/dichloroethene, 271, 274
schematic of tracer injection apparatus, 272*f*
strong elution tailing and rebound, 280
tracer injection method, 271
trichloroethene and dichloroethene data, 276–277
vertical distribution of trichloroethene within aquifer, 278*f*
vertical spatial variability of hydraulic conductivity and contaminant concentrations within aquifer, 277, 280

Aeration, in-well. *See* Enhanced remediation demonstrations at Hill AFB; In-well aeration for remediation of contaminated aquifer

Air sparging with soil vapor extraction
air sparging (AS) technique, 153
concentrations and masses of target compounds from soil sample analysis, 160*t*
concentrations of target compounds in groundwater analysis and mass removal estimates for pump-and-treat based on partitioning tracer tests, 164*t*
concentrations of three target compounds in soil vapor extraction (SVE) offgas during SVE and AS/SVE treatment, 162*f*
contamination in vadose and saturated zones, 156
field experiment location, 155–156
field site description, 154–155
groundwater analyses before and after treatment, 163–165
masses of constituent compounds collected in SVE offgas, 163*t*
monitoring CO_2 in offgas, 163

285

column test of soil cores from coal tar site and laboratory-packed columns, 229
comparison of saturation estimates from tracer method and soxhlet extractions for laboratory-packed columns and field cores, 234, 235*f*
field tracer test method, 231
field tracer tests agreeing with laboratory partitioning tracer tests, 236
laboratory column studies, 233–236
method of moments determining coal tar saturation, 236
partition coefficient determination method, 228–229
partitioning interwell tracer test (PITT) method, 227–228
PITT sampling larger area than traditional methods, 228
retardation of partitioning tracer, 227–228
schematic of field cores and approximate depths, 230*f*
useful method for characterization of coal tar saturation, 236, 238
Performance test, factors for conducting proper, 4–5
Permeable reactive barrier (PRB) with zero-valent metal. *See* Groundwater remediation of chromium
Phase distribution of contaminants
partitioning between groundwater and sediments, 199–200
See also Evaluation sampling plans for innovative remediation technology
Pilot treatment evaluation
alternative methods of contaminant mass estimation, 203–206
example of subsurface variability, 201–203
pre-treatment and post-treatment sediment-associated carbon tetrachloride measurements, 202*f*, 203*f*
See also Evaluation sampling plans for innovative remediation technology
Point sampling methods
coring for solid-phase sampling, 4
groundwater monitoring wells, 4
technologies based on geoprobes, 4
Publication, dissemination of results, 5
Pump-and-treat approach
baseline for groundwater extraction well monitoring during partitioning tracer tests, 165
conventional method of treating DNAPL-contaminated aquifers, 65

remediating groundwater contamination, 2–3
See also Surfactant enhanced subsurface remediation; Water-flush (WF)

R

Radio frequency heating
areas for future development, 28
common power delivery methods to soil, 26
dielectric heating, 27–28
electromagnetic technique for recovery of organics, 25–28
major technique limitations, 27
pilot-scale remediation tests in field, 26
surface equipment, 27
See also Soil heating
Rapid fire analyses
role in mass estimation, 199–200
See also Evaluation sampling plans for innovative remediation technology
Remediation demonstrations. *See* Enhanced remediation demonstrations at Hill AFB
Remediation of groundwater contamination
challenge of subsurface complexities, 4
need for field demonstrations of innovative technologies, 4–5
need for innovative technologies, 2–4
Remediation technology. *See* Evaluation sampling plans for innovative remediation technology
Resistive heating
field deployable stage of testing, 29–30
internal heating technique passing AC current through soil, 29–30
limitation to applicability in permeable sediments, 30
most applicable to vadose zone, 30
primary research areas, 30
surface equipment variations, 30
See also Soil heating

S

Sampling plans for remediation technology. *See* Evaluation sampling plans for innovative remediation technology
Saturated zone remediation
electro-osmosis technique, 32
microwave heating, 28
pump-and-treat ineffective for NAPL removal, 136–137
See also Cyclodextrin for enhanced in situ flushing; Water-flush (WF)